# SpringerBriefs in Petroleum Geoscience & Engineering

**Series Editors**

Dorrik Stow, Institute of Petroleum Engineering, Heriot-Watt University, Edinburgh, UK

Mark Bentley, AGR TRACS International Ltd, Aberdeen, UK

Jebraeel Gholinezhad, School of Engineering, University of Portsmouth, Portsmouth, UK

Lateef Akanji, Petroleum Engineering, University of Aberdeen, Aberdeen, UK

Khalik Mohamad Sabil, School of Energy, Geoscience, Infrastructure and Society, Heriot-Watt University, Edinburgh, UK

Susan Agar, Oil & Energy, Aramco Research Center, Houston, USA

Kenichi Soga, Department of Civil and Environmental Engineering, University of California, Berkeley, USA

A. A. Sulaimon, Department of Petroleum Engineering, Universiti Teknologi PETRONAS, Seri Iskandar, Malaysia

The SpringerBriefs series in Petroleum Geoscience & Engineering promotes and expedites the dissemination of substantive new research results, state-of-the-art subject reviews and tutorial overviews in the field of petroleum exploration, petroleum engineering and production technology. The subject focus is on upstream exploration and production, subsurface geoscience and engineering. These concise summaries (50-125 pages) will include cutting-edge research, analytical methods, advanced modelling techniques and practical applications. Coverage will extend to all theoretical and applied aspects of the field, including traditional drilling, shale-gas fracking, deepwater sedimentology, seismic exploration, pore-flow modelling and petroleum economics. Topics include but are not limited to:

- Petroleum Geology & Geophysics
- Exploration: Conventional and Unconventional
- Seismic Interpretation
- Formation Evaluation (well logging)
- Drilling and Completion
- Hydraulic Fracturing
- Geomechanics
- Reservoir Simulation and Modelling
- Flow in Porous Media: from nano- to field-scale
- Reservoir Engineering
- Production Engineering
- Well Engineering; Design, Decommissioning and Abandonment
- Petroleum Systems; Instrumentation and Control
- Flow Assurance, Mineral Scale & Hydrates
- Reservoir and Well Intervention
- Reservoir Stimulation
- Oilfield Chemistry
- Risk and Uncertainty
- Petroleum Economics and Energy Policy

Contributions to the series can be made by submitting a proposal to the responsible Springer contact, Anthony Doyle at anthony.doyle@springer.com, or the Academic Series Editor, Prof Dorrik Stow at dorrik.stow@pet.hw.ac.uk.

More information about this series at http://www.springer.com/series/15391

Shuvajit Bhattacharya

# A Primer on Machine Learning in Subsurface Geosciences

 Springer

Shuvajit Bhattacharya
Bureau of Economic Geology
The University of Texas at Austin
Austin, TX, USA

ISSN 2509-3126           ISSN 2509-3134  (electronic)
SpringerBriefs in Petroleum Geoscience & Engineering
ISBN 978-3-030-71767-4           ISBN 978-3-030-71768-1  (eBook)
https://doi.org/10.1007/978-3-030-71768-1

This Springer imprint is published by the registered company Springer Nature Switzerland AG
The registered company address is: Gewerbestrasse 11, 6330 Cham, Switzerland

# Preface

The application of traditional machine learning and emerging deep learning algorithms in subsurface geosciences is now a hot topic. The advent of big data analytics is changing the conventional workflows used in the subsurface community at various levels. Many new organizations, irrespective of industry and academia, are adopting data analytics and machine learning for the first time. Today's geoscientists are eager to learn new techniques and methods in data analytics to solve their geoscience problems.

This book provides with a concise review of data analytics and popular machine learning algorithms and their applications in subsurface geosciences, specifically geology, geophysics, and petrophysics. Machine learning is a part of data analytics. I emphasize machine learning in this book, concisely.

This book was written to aid other machine learning practitioners and newbies—including students—in geodata analytics. This book is intended for geoscientists and reservoir engineers of various specialties. In this book, I attempt to impart a basic understanding of data analytics (DA) and machine learning (ML) and how we can use these tools to solve our problems more efficiently and consistently, regardless of programming language.

Language is often a problem when it comes to new techniques and methods. ML is no exception. There are hundreds of terms and abbreviations commonly used in the computer science and engineering communities that are unfamiliar to geoscientists. I have tried to use language familiar to geoscientists while gently introducing these new concepts. This book will not turn geoscientists into programmers overnight, but it will help them understand the fundamentals of ML and how to apply these methods to geoscience data.

This book provides a timely review and discussion of the fundamentals, workflow, proven success, promises, and perils of ML. It can be used as a ready-to-go reference for understanding machine learning and its nuances in both subsurface and surface applications. This book will provide necessary knowledge regarding:

1. the existing approaches in exploratory geoscience data analysis and their limitations,

2. fundamental concepts and applications of traditional ML and emerging deep learning algorithms,
3. necessary steps in ML model development,
4. identification of fit-for-purpose ML algorithms for real-world problems, and
5. the future of ML in geosciences.

Chapter 1 deals with the necessary foundations of data analytics, machine learning, geoscience databases, and the concept of scales. An overview of the history of ML narrates how different algorithms came into the community, offered new solutions, and were supplanted by new algorithms offering better new solutions.

Chapter 2 systematically discusses different statistical measures used in the geosciences and provides examples. This will prepare readers to understand the types of data analytics applied to geoscience data.

Chapter 3 deals with the basic ML workflow, including supervised, unsupervised, and semi-supervised approaches. The concepts of deep learning workflow are also included.

Chapter 4 provides a brief review of popular ML algorithms, including emerging deep learning and physics-informed ML. Each algorithm is covered with its fundamentals, network hyperparameters, optimization, and geoscience-specific examples.

Chapter 5 summarizes various ML applications in structure, stratigraphy, rock properties, and fluid-flow analysis using core, well log, seismic, and fiber-optic data.

Chapter 6 discusses the differences in modern data analytics approaches from past approaches, current challenges, and opportunities for geoscientists in both industry and academia. I also lay out a few specific research directions that ML practitioners in geosciences may want to engage in the next several years.

One of the points that I reiterate in this book is that the failure of a data analytics project is not always with the algorithms themselves, but rather our lack of understanding of the operational processes, underlying assumptions, and limits of those algorithms, as well as our datasets. In this book, I try to bring you back to the basics, explain geoscience databases with an ML spin, and explore the areas in geoscience that are ready for ML applications. I hope this book will help popularize ML in the subsurface community and among those interested in energy resources, water, and fluid storage-related projects. I hope you will enjoy reading this book and solving your own problems.

Austin, USA                                                    Shuvajit Bhattacharya

# Acknowledgments

I would like to express my heartfelt thanks to many individuals from whom I personally learned a great deal about data analytics and quantitative aspects of geosciences: Dr. Tim Carr, Dr. Kurt Marfurt, Dr. Mahesh Pal, Dr. Shahab Mohaghegh, and Dr. Srikanta Mishra. Thanks to my colleagues and supervisors at the University of Alaska Anchorage and the University of Texas at Austin for their support and encouragement. The preparation of this manuscript was partly supported by a publication grant from the Bureau of Economic Geology at the University of Texas at Austin. Thanks to the inventors of machine learning algorithms, developers of R, Python, Julia, scikit-learn, and tensorflow, and the writers of informative blog posts (e.g., toward data science and Machine Learning Mastery). I would also like to express thanks to numerous individuals and professionals with whom I had an opportunity to discuss artificial intelligence and learn from their experiences.

Special thanks go to Dr. Anthony Doyle and Ashok Arumairaj at Springer Nature for making this book happen. Thanks to Dr. Shayan Tavassoli and Emily Harris at the Bureau of Economic Geology for proofreading this book. I also acknowledge all the publishers and individuals who provided permissions to use figures from their technical articles and websites. I deeply appreciate all your support and encouragement.

Shuvajit Bhattacharya

# Contents

# About the Author

**Dr. Shuvajit Bhattacharya** is a researcher at the Bureau of Economic Geology, the University of Texas at Austin. He is an applied geophysicist/petrophysicist specializing in seismic interpretation, petrophysical modeling, machine learning, and integrated subsurface characterization. He uses advanced computational technologies to solve complex problems in geosciences, which have societal and economic impacts. Dr. Bhattacharya has completed several projects in diverse geologic settings in the US, Norway, Australia, South Africa, and India. He has worked in both academia and industry. He has published and presented more than 50 technical articles in peer-reviewed journals and conferences. His current research focuses on the pressing issues and frontier technologies in energy exploration, development, and subsurface fluid storage (carbon, hydrogen, and wastewater). He completed his Ph.D. at West Virginia University and M.Sc. at the Indian Institute of Technology Bombay.

# Acronyms

| | |
|---|---|
| AAPG | American Association of Petroleum Geologists |
| AI | Artificial Intelligence |
| AIC | Akaike Information Criterion |
| ANFIS | Adaptive Network-Based Fuzzy Inference System |
| ANN | Artificial Neural Network |
| AR | Augmented Reality |
| BA | Bat-Inspired Algorithm |
| BDeu | Bayes Birichlet Likelihood-Equivalence Uniform Joint Distribution |
| BI | Brittleness Index |
| BN | Bayesian Network |
| BPNN | Backpropagation Neural Network |
| CGR | Computed Gamma-Ray |
| CM | Committee Machine |
| CNN | Convolutional Neural Network |
| CT | Computed Tomography |
| CV | Cross-Validation |
| DA | Data Analytics |
| DAG | Directed Acyclic Graphs |
| DAS | Distributed Acoustic Sensing |
| DBSCAN | Density-Based Spatial Clustering of Applications with Noise |
| DL | Deep Learning |
| DNN | Deep Neural Network |
| DT | Decision Tree |
| DTS | Distributed Temperature Sensing |
| FCN | Fully Connected Network |
| GA | Genetic Algorithm |
| GAN | Generative Adversarial Network |
| GBM | Gradient-Boosting Machine |
| GLCM | Gray-Level Co-Occurrence Matrix |
| GPU | Graphical Processing Unit |
| GR | Gamma-Ray |

| | |
|---|---|
| GTM | Generative Topographic Mapping |
| HCA | Hierarchical Cluster Analysis |
| ICA | Independent Component Analysis |
| IDLM | Integrated Deep Learning Model |
| IP | Initial Production |
| KNN | K-Nearest Neighbor |
| KPI | Key Performance Indicators |
| LIME | Local Interpretable Model-Agnostic Explanations |
| LSTM | Long Short-Term Memory |
| MAE | Mean Absolute Error |
| MBE | Mean Bias Error |
| MDA | Mean Decrease Accuracy |
| MDI | Mean Decrease Impurity |
| MDL | Minimum Description Length |
| ML | Machine Learning |
| MLFN | Multi-Layer Feed-Forward Neural Network |
| MLP | Multi-Layer Perceptron |
| MLPNN | Multi-Layer Perceptron Neural Network |
| MRGC | Multi-Resolution Graph-Based Clustering |
| MSE | Mean-Squared Error |
| NMR | Nuclear Magnetic Resonance |
| NPHI | Neutron Porosity |
| PCA | Principal Component Analysis |
| PDP | Partial Dependence Plots |
| PEF | Photo-Electric Factor |
| PNN | Probabilistic Neural Network |
| PSO | Particle Swarm Optimization |
| QA | Quality Assurance |
| QC | Quality Check |
| RBF | Radial Basis Function |
| ReLu | Rectified Linear Unit |
| RF | Random Forest |
| RGB | Red Green Blue |
| RHOB | Bulk Density |
| RHOmaa | Apparent Matrix Grain Density |
| RMS | Root Mean Square |
| RMSE | Root Mean Square Error |
| RNN | Recurrent Neural Network |
| SEG | Society of Exploration Geophysicists |
| SEM | Scanning Electron Microscopy |
| SEPM | Society for Sedimentary Geology |
| SGD | Stochastic Gradient Descent |
| SHAP | SHapley Additive exPlanations |
| SMOTE | Synthetic Minority Over-Sampling Technique |

| | |
|---|---|
| SOM | Self-Organizing Map |
| SPWLA | Society of Petrophysicists and Well Log Analysts |
| SVM | Support Vector Machine |
| SVR | Support Vector Regression |
| TICC | Toeplitz Inverse Covariance Clustering |
| TOB | Transparent Open Box |
| TOC | Total Organic Carbon |
| Umaa | Apparent Matrix Volumetric Cross Section |
| VR | Virtual Reality |
| XGBoost | Extreme Gradient Boosting |
| XRD | X-Ray Diffraction |
| XRF | X-Ray Fluorescence |

# Chapter 1
# Introduction

**Abstract** In the first chapter, we will learn about big data, data analytics, and machine learning, as well as their utilities in different disciplines, including geosciences. A thorough understanding of data analytics problems, algorithms, and geoscience-specific data is critical before applying these sophisticated tools. We will also go into the brief history of the advent of different machine learning algorithms, the different types of geoscience databases, and their fitness to machine learning applications.

**Keywords** Big data · Data analytics · Machine learning · Data analytics types · Geoscience databases · Nature of geoscience data

## 1.1 What are Big Data, Data Analytics, and Machine Learning?

### 1.1.1 Big Data

As of 2021, we are experiencing an unprecedented technological advancement and transformation in terms of scale, scope, complexity, and global impact. A range of new technologies is fusing the physical, digital, and biological worlds as never before. Some thought-leaders are referring to this period as the fourth industrial revolution, or Industry 4.0. The prior three industrial revolutions involved the onset of mechanization at the end of the eighteenth century (Industry 1.0); the emergence of electricity, gas, and oil at the end of the nineteenth century (Industry 2.0); and the invention of computers and electronics in the second half of the twentieth century (Industry 3.0). The wave of fourth revolution that we are currently riding is underpinned by big data, digitilization, and data analytics. As Google's Eric Schmidt said, "From the dawn of civilization until 2003, humankind generated five exabytes of data. Now we produce five exabytes every two days…and the pace is accelerating."

There is a myth about what big data is. The definition of big data has changed over the years with the advent of greater computational power such as the graphics

© The Author(s), under exclusive license to Springer Nature Switzerland AG 2021
S. Bhattacharya, *A Primer on Machine Learning in Subsurface Geosciences*,
SpringerBriefs in Petroleum Geoscience & Engineering,
https://doi.org/10.1007/978-3-030-71768-1_1

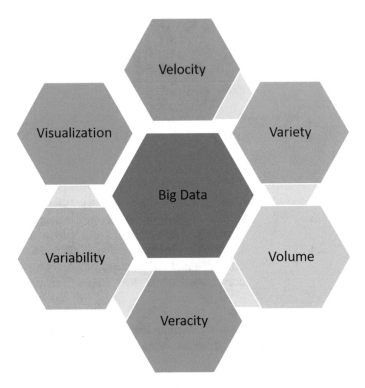

**Fig. 1.1** The multiple dimensions that define **b**ig **d**ata

processing unit (GPU). Simply put, big data is any data that we cannot analyze on our personal computers. We often describe big data with three v's: *velocity, variety,* and *volume*. Velocity represents real-time data (e.g., fiber-optic data, logging-while-drilling data, and fluid production). Variety represents the data coming from different domains and scales (e.g., core, well logs, seismic, drilling, and completions). Volume indicates the size of the data that cannot be handled using personal computers (Bhattacharya et al. 2019). The term *veracity* is also used to indicate that certain data comes with uncertainties. Figure 1.1 shows the simplified concept of big data. Recently, data scientists have also added *variability* and *visualization* to the definition of big data.

### 1.1.2 Data Analytics

Data is at the heart of data analytics. Data analytics is the science of examining data to identify trends and draw conclusions from them, which we can use to make actionable decisions. It deals with fundamental principles, methods, processes, and techniques to provide hindsight, insight, and forecasts from the available data. Conway (2010) depicts data science in his famous Venn diagram, showing the intersection of math and stat knowledge, domain expertise, and also hacking skills (!), which attempts to capture some of the essential skills needed in data science. Humor aside, data scientists must have a solid foundation in mathematics, statistics, and domain expertise, though the exact mixture may vary depending on the role and business application. In addition to these major skills, data scientists must be able to understand business problems, derive value from data analytics solutions, and communicate their conclusions effectively.

Data science is a diverse, multidisciplinary field. It is an emerging field and currently one of the hottest job sectors. Data scientists are employed by organizations dealing with finance, retail, marketing, health care, information, energy, manufacturing, and scientific and technical services. A 2017 report by the European Commission projected that the number of professionals in this field would increase to 10.43 million with a compound average growth rate of 14.1% by 2020. The U.S. Bureau of Labor Statistics projects there will be about 11.5 million new jobs in data science by 2026.

There are different components of data analytics that we need to understand. Analytics is not a one-time study that we conduct, present, and then put back to the shelf. It is an ongoing process. We need to be mindful of certain data analytics components if we really want to harness big data and build an excellent data-centric strategy for our businesses. These key components are reliable upon data acquisition sources, standard programs to analyze data, data security, standardized data governance, data migration, data storage, data processing, data visualization, data integration, data analysis, optimization, knowledge discovery, and data ethics. Although these terms may sound more like Information Technology, we as geoscientists have responsibilities to adopt these best practices, whether in the industry, research labs, or academia. For example, we would want to use an up-to-date and consistent coordinate system for the same data across several projects. We would also like to use the same unit for a particular geophysical measurement (such as sonic velocity and neutron porosity). The same thing goes for geodata file formats. Different formats are used for the same type of data. In such cases, the upper management (or at least the project managers at the group level) should develop an objective-oriented operational model consistent across the units to make these things standardized, well-documented, and the enabler of employee success. Only then will we start receiving the dividends back from our investment in data analytics.

### 1.1.3 Machine Learning

Machine learning (ML) is part of data analytics. In this book, I emphasize machine learning. Tom Mitchell, the renowned Carnegie Mellon professor, defined ML as "the study of computer algorithms that allow computer programs to automatically improve through experience" (1997). He went on to formalize the definition of ML: "A computer program is said to learn from experience $E$ with respect to some class of tasks $T$ and performance measure $P$, if its performance at tasks in $T$, as measured

**Fig. 1.2**  ETP definition of machine learning (after Mitchell 1997)

by $P$, improves with experience $E$" (Fig. 1.2). This means that ML can improve the performance of executing a task over time. We do not need to memorize this formal definition, but understand it, as ML is more about the practice and yielding knowledge from data-driven modeling. These tasks can vary and may include clustering, classification, regression, outlier detection, etc. We use metrics such as accuracy, false positive, false negative, and errors to measure model performance. We run these models using supervised, semi-supervised, and reinforcement learning approaches (see Chapter 3 for more details).

## 1.2  History of Machine Learning

ML may seem a buzzword these days; however, science fiction reveals that human beings were always fascinated by the idea of creating machines having human-like qualities. The mathematical formulation of ML began in the early 1940's (Fig. 1.3). The seminal paper by Warren McCulloch and Walter Pitts (1943) is often cited as the starting point of modern ML. Their work provided the first logic-based theory on the mind and the brain, building on Alan Turing's notion of 'automatic machines' first described in 1936. McCulloch and Pitts were motivated to find how human brains learn (biological learning). Their simple model was fed by many binary inputs (or neurons), which were then processed to generate a binary output (i.e., ones and zeros). This simple model was later modified by combining several neurons to generate complex functions that could learn non-linear behavior.

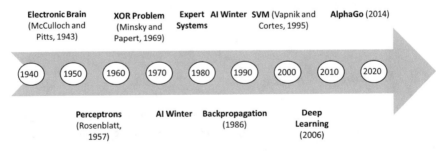

**Fig. 1.3**  A simplified timeline of major milestones in AI research since the 1940's

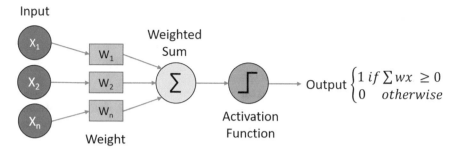

**Fig. 1.4** A simplified diagram of a perceptron model, consisting of input, weight, activation function, and output. Activation functions will generate an output of one when a certain threshold is crossed; otherwise, it will provide zero as the output

In 1958, Frank Rosenblatt introduced the concept of a single-layer perceptron, or "the first machine which is capable of having an original idea" (artificial intelligence, or AI). This research, funded by the U.S. Office of Naval Research, demonstrated that an IBM 704 computer (which was the size of a room) could learn to distinguish punch cards marked on the left from the cards marked on the right after 50 trials. This is a classic example of supervised learning. Remarkably, this perceptron model would accept different weights from different inputs, meaning different inputs have different levels of impact, which is more realistic (Fig. 1.4). The contemporary media went gaga over this invention. Headlines read "NEW NAVY DEVICE LEARNS BY DOING: Psychologist Shows Embryo of Computer Designed to Read and Grow Wiser." Although the single-layer perceptron model showed tremendous promise at that time, critics were concerned about its global generalization ability (Minsky and Selfridge 1961; Minsky and Papert 1969). Some of these studies cast serious doubts on the perceptron model and artificial intelligence in general. Since then, artificial intelligence has experienced several winters, depending on the advent of computational technologies, trust in the models, funding, and data.

Neural networks blossomed again in 1986 when Rumelhart, Hinton, and Williams formally introduced and popularized the concept of back-propagation. Earlier, Paul Werbos had proposed that back-propagation could be used for a neural network in his 1974 Ph.D. thesis. Back-propagation in a network repeatedly adjusts its weights to neuron connections to minimize the difference between the actual output and the desired output. The back-propagation network was widely utilized in the AI community then and still used today. Unfortunately, the AI community soon faced a dip in interest and funding (the second AI winter), perhaps due to the lack of adequate computational power and the capabilities of neural network tools being oversold.

In 1989, works by Hornik et al. on a multilayer feedforward neural network and by LeCun on back-propagation in recognizing handwritten zip codes provided new research directions in machine learning. The method proposed by LeCun went on to be used by banks to read checks. In 1995, Vapnik and Cortes introduced their milestone paper on kernel-based support vector machines (SVMs), which were very different from a black box–like multilayer perceptron neural network. This line of independent research and development of novel ML algorithms was much in need at the time to keep the field of ML alive and unravel its complexities. Based on a solid foundation of statistical learning theory and mathematics, SVMs provided a great opportunity for early naysayers of ML to get their feet wet and apply ML in different domains, which further popularized this technique.

Other algorithms, such as recurrent neural networks and random forest, were invented between 1980 and 2000. In these two decades, the popularization of video games and the application of graphics processing units (GPUs)—previously used only in video games—toward more general computing purposes also helped to advance machine learning. Breiman's random forest (2001) gained much popularity because of its simple, tree-like structure that was easy to understand and implement to datasets. As we will later find out, each of these algorithms has its strengths and weaknesses, and some can outperform others in various problems.

The third wave of AI arrived in 2006, when Geoffrey Hinton came up with breakthrough research on greedy layer-wise training to train a deep belief network. This work is often credited as popularizing deep learning. In 2014, Goodfellow et al. released their generative adversarial network (GAN), in which two neural networks compete against each other. GAN has gained significant popularity in image analysis now. Several new network structures, such as UNet, SegNet, and ImageNet were launched in 2015 and 2016.

Deep learning's superior performance in image pattern recognition and natural language processing over traditional machine learning algorithms became apparent, leading to a meteoric rise in further research and deployment of deep learning. According to Manning, the deep learning tsunami hit the AI community in 2015. Since then, several AI startups have been born, with some acquired by large companies. In 2015, Facebook deployed DeepFace, a feature based on deep learning that could automatically tag and identify its users in photographs. In 2016, Google DeepMind's algorithm AlphaGo beat professional Go player Lee Sedol at a highly publicized tournament. These breakthroughs in AI over the years were the culmination of dedicated and tireless efforts by universities, research labs, industry, and students. In several cases, student interns went on to develop better solutions than the existing ones. At Microsoft, George Dahl and Abdel-rahman Mohamed developed work on the use of GPUs for automated speech recognition. At Google, Navdeep Jaitly's work on deep learning improved Android's speech recognition power (Jaitly et al. 2012).

In late 2010's, the concepts of programming and ML became far more accessible with a wave of affordable and readily available online courses from providers like Coursera, Udemy, and Udacity. Online documentation such as publicly available code and data, informative blogposts (e.g., towardsdatascience), and GitHub repositories further enabled knowledge sharing in the AI/ML communities. Advancements in computers and mobile workstations made it possible to crunch big data without expensive or specialized equipment. All these factors radically changed the way we learn and do research. Many refer to this as the democratization of AI, which has fostered new research and deployment of AI across many disciplines. Hopefully, there will be more and more research in this domain in the future. We should also be mindful of the perils of such research, including fake images and videos generated using AI. We all must learn about and practice proper ethics and integrity in data analytics.

## 1.3  Where are the Geoscientists in this Digital Age and ML-Tsunami?

Similar to ML practitioners in many other disciplines, geoscientists are learning these new ML technologies and terminologies and adopting them for geosciences as needed. This is because many of the concepts widely used in geosciences can be formulated mathematically and analyzed in a better quantitative manner. For example, Markov chains can be used for facies characterization (Krumbein and Dacey 1969; Carr 1982). Krumbein is often cited as the father of mathematical geology. Krumbein started implementing some fundamental statistical measurements to sedimentology. Despite the potential for wide-ranging applications, geoscience has been slow to adopt new ML technologies. There are several reasons behind this, including the seemingly descriptive nature of the discipline, multi-scale heterogeneities, rare events, missing time, challenges in converting some process-based cognitive geologic concepts to mathematical forms, ill-posedness of problems (i.e., more output than the available input), and the lack of labeled large database open to the public. In addition, most subsurface data are sampled in a highly biased manner, sparse, and noisy. The cost of subsurface data acquisition, processing, and maintenance is another obstacle. For example, 3D seismic and downhole fiber-optic data cost millions of dollars for just a small area.

Despite the broad challenges, a few specialties in geosciences have been implementing ML in their work since the 1990's, notably remote sensing and geophysics. Remote sensing has been an early adopter of data-driven modeling and ML (Atkinson and Tatnall 1997; Pal and Mather 2003). Several first works in subsurface geosciences have focused on dimensionality reduction, clustering, and supervised classification using principal component analysis, discriminant analysis, and artificial neural networks (ANNs) (McCormack 1991; Doveton 1994; Luthi and Bryant 1997). Since the 2000's, there has been more applications of different ML algorithms (i.e., ANNs, SVMs, fuzzy logic, and genetic algorithms) in the geosciences, including geophysics, petrophysics, and hydrology (Govindarazu and Rao 2000; Kuzma 2003; Li and Castagna 2004; Qi and Carr 2006).

The application of AI has virtually exploded in the subsurface geoscience community since 2015–2016. Before 2015, the number of articles published in the American Association of Petroleum Geologists (AAPG) and Society of Exploration Geophysicists (SEG) annual meetings was meager. Coincidentally, this was before the historic downfall in oil price in 2015. This trend suddenly changed after 2016, when several new technical sessions and workshops were introduced in these annual meetings and their regional equivalents. Special interest groups were formed in some of these societies, such as AAPG and SPWLA, to share knowledge and foster collaboration across the industry and academia. A few of these professional societies and organizations—for example, SEG, TGS-Kaggle, and FORCE—also started organizing hackathons on open datasets since 2016, in which the competitors had to release their codes. This occurred in tandem with the industry and geological surveys releasing massive subsurface datasets (e.g., seismic, well log, core, and production data) to the public. Of course, many of these sudden changes were introduced in a top–down management approach in the industry, which was needed to break the status quo and make the businesses more efficient, collaborative, and productive with limited resources. Currently, data analytics and machine learning are being applied in geophysics, petrophysics, hydrology, structure, stratigraphy, geochemistry, and paleoclimate studies (Li and Misra 2017; Waldeland and Solberg 2017; Araya-Polo et al. 2017; Ross et al. 2018; Scher 2018; Wu et al. 2019; Zuo et al. 2019).

## 1.4   Why should we care about Machine Learning in Geosciences?

Although ML is relatively new to geosciences, we must learn how to apply it to our problems appropriately. At a high-level, this new tool will enable us to think more analytically, solve problems consistently and quantitatively, and further transdisciplinary collaboration. ML can help us in several areas of geosciences, including tectonics, stratigraphy, geophysics, geochemistry, petrophysics, hydrology, paleoclimate, paleontology, remote sensing, and planetary geology. ML will also help us build data-driven models rather than purely conceptual models that are hard to verify with real data. Specifically, ML can assist us in the following tasks, represented by Fig. 1.5:

1.   Outlier detection: We can use ML to go over event logs and other data points to detect anomalous patterns in the database. Once outliers are detected, geoscientists can use domain knowledge to assess whether these are true outliers (i.e., bad data records) or indicative of certain geologic phenomena (i.e., special lithology or fluid). This is very useful for geophysical data such as seismic and well logs. This will also help geoscientists in properly normalizing data without removing geologic trends.
2.   Transcription: Although there has not yet been widespread use of ML for data transcription in geosciences, the technology holds massive potential. For example, large, unstructured geodata can be converted into a more structured format for digital archiving. This application is of interest to industries, libraries, museums, and state and national agencies.
3.   Structural annotation: Often, geodata are not labeled or only partially labeled. Using ML, geoscientists can annotate (or add metadata to) images based on their

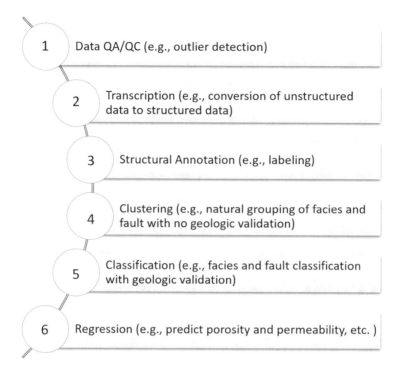

**Fig. 1.5** A typical set of geoscience tasks (with examples) that can be performed using machine learning

location, type, geometric relation, etc. This is especially applicable to seismic, outcrop, and remote sensing images covering large areas.

4. Clustering: We can use ML to form groups based on natural data distribution using the concept of unsupervised learning. This provides a rudimentary understanding of data relationships and help geoscientists form working hypotheses to test. Once ML-based groups are calibrated with the ground-truth data, they can be used in supervised and semi-supervised ML models.

5. Classification: We can use semi-supervised and supervised ML to classify objects such as facies, faults, fractures, pore spaces, and certain minerals. We can do this using a database composed of either numerical values (e.g., seismic and well logs) or images (e.g., thin sections, outcrops, and remote sensing).

6. Regression: We can implement ML to predict features of interest, such as porosity, permeability, fluid production, or even missing sample records. These problems are common in geosciences because we often do not have access to the required sensors or are interested in forecasting certain parameters in real time.

The application of ML in real-time would also provide significant value to geoscientists, helping especially those working on projects in energy and water resources, storage, and geohazards to make real-time decisions on field operations. Several algorithms (such as long short-term memory) are suited for such time series analysis in a predictive way. Ultimately, we should remember that ML is like a double-edged sword. If we use it carefully with proper training, we can gain many wonderful things; otherwise, there are ample chances for disasters.

## 1.5   Types of Data Analytics

Gartner (2012) divided data analytics into four different phases based on the relationship between difficulty and derived value (Fig. 1.6). These four phases include: descriptive analytics, diagnostic analytics, predictive analytics, and prescriptive analytics. Understanding each of these types is helpful throughout the lifecycle of a data analytics project.

Descriptive analytics is the first step in data analytics. In this step, we analyze both historical and real-time existing data. Simple statistical analyses and visualization are useful at this stage, which is more of a standard reporting, dashboard, and drill-down type exercise. As scientists, we are more interested in causality, not just the correlation between certain existing variables.

In diagnostic analytics, we analyze why certain variables show certain relationships. In this step, we must use our domain expertise to analyze the data and find the reasons behind certain behavior. This helps us discover knowledge and select meaningful attributes for the next stage of predictive analytics. A few sophisticated techniques, such as the Bayesian network, can be useful to infer the direction of causality.

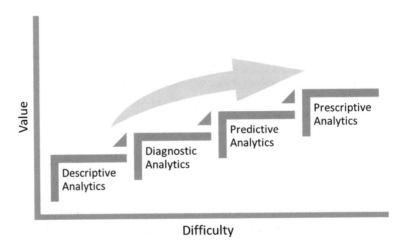

**Fig. 1.6**  The four stages of data analytics (after Gartner 2012). If everything goes well with the final stage, prescriptive analytics, we can apply specific knowledge derived from the previous phases of the ongoing study to another area (analog or asset) and recommend certain types of data acquisition and workflow adoption

The next step is predictive analytics. This stage is about understanding the past to predict the future. In this step, we apply traditional ML and deep learning algorithms to predict features of interest (e.g., facies, fractures, porosity, permeability, fluid production, carbon price, etc.) using historical and real-time data. For predictive analytics, we need a large volume of data. Sometimes, we do not have enough data, especially in the case of image analysis. In those cases, we can generate synthetic data based on certain rules and domain knowledge. We can add white noise to such data to make it more like real-world data, a method often employed in geophysics.

Prescriptive analytics is the last step in data analytics. At this stage, we have already performed several experiments, modeling, and sensitivity analysis that have provided critical insights into the data, their relationships, and the underlying rules (e.g., geology, physics, and chemistry). This stage is based on optimization techniques. Prescriptive analytics help businesses generate multiple scenarios, forecast the possible outcomes in each of these scenarios, and recommend the best possible action. Keep in mind that prescriptive analytics is more complicated than other analytics approaches to administer. When implemented correctly, it can make businesses or research more productive and efficient.

## 1.6 Geoscience Databases

Data in geosciences come in all types and amounts, from a handful of lab analyses to petabytes of remotely sensed subsurface and satellite data. Depending on the mode of data acquisition, tools, and project objectives, we deal with a variety of data in geosciences, including numerical and image data. However, these different types of data are also present in other disciplines, not just geosciences.

### 1.6.1 Numerical Data Types

Numerical data in geosciences can be divided into two broad types: discrete and continuous (Wessel 2007). Discrete variables have distinct integer values (i.e., a finite number), whereas continuous variables can take any possible values without any breaks (Fig. 1.7). For example, the number of times we flip a coin is a discrete variable. Length is a continuous variable, as it can take any possible values.

#### 1.6.1.1 Discrete Variables

There are three types of discrete variables: count, ordinal, and nominal.

Count refers to the number of experiments or samples. Examples include the number of fossils in an area, the number of pyrite framboids in a scanning electron microscopy (SEM) image, and the number of fractures in a unit area.

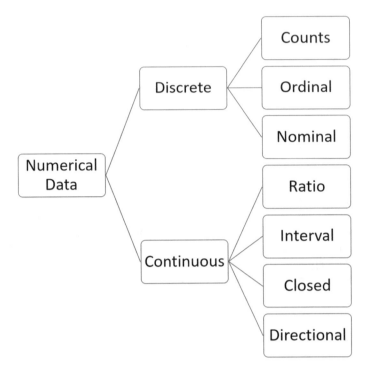

**Fig. 1.7**  Various types of numerical data commonly used in geosciences

Ordinal variables refer to ranking with respect to a certain reference. Examples of ordinal variables in geosciences include the Moh's hardness scale, in which each consecutive mineral is harder than the previous one. Another example is the Mercalli scale of earthquake intensity. In this intensity scale, relative perceptions of earthquake intensity among humans are used to infer earthquake strength. Although it has certain relations to the Richter scale, the Mercalli scale is not valuable to seismologists and geotechnical engineers.

Nominal, or categorical, variables are prevalent in geosciences. In these cases, a name can be converted to a numerical format for graphical plotting and analyses, similar to Boolean algebra. We can do it either by integer encoding (e.g., 1, 2, 3,.., n) or one-hot encoding (e.g., 1 or 0). In the ML community, one-hot encoding is more common now. These new variables are also called dummy variables. For example, consider the presence of facies, fractures, and fossils in rocks. If certain facies are present, we can assign them a value of one; if not, we can assign them zero. This way, we can build a time series showing the presence or absence of a particular facies in the stratigraphic record. Because we deal with several facies in a stratigraphic interval, we can convert each facies to a particular numerical value. We can easily transform the text to a numerical value by using a simple if–then-else statement in a program. This is important for ML applications and will be covered extensively in the data preparation stage (Chapter 3). Unlike facies, simple fault classification problems are binary problems.

### 1.6.1.2 Continuous Variables

Continuous variables can be divided into at least four types: ratio, interval, closed, and directional.

Ratio data have a fixed zero value as the starting point. Examples include age, length, width, and mass. Many rock properties from well logs and seismic data are ratio data, such as resistivity, density, photoelectric factor, and velocity.

Interval variables differ from ratio variables in that the zero value in these variables is not the end of the scale. For example, temperature in Celsius or Fahrenheit is interval data because negative temperature values are possible. However, the same temperature data in Kelvin is ratio data. Also, keep in mind that the negative values in certain wireline log responses (e.g., neutron porosity and density porosity) do not mean porosity is negative. Rather, it means the formations are tight. Porosity is neither a ratio nor an interval variable; it is a closed variable.

Closed variables are described in the form of percentages and ratios (i.e., parts per million, etc.). The sum of the closed variables equals one or 100, implying a closed system or universe. Geochemists and geophysicists frequently use the concept of closed variables. For example, the proportions of quartz, clay, and carbonate in a rock are closed variables. We often use ternary diagrams to plot such data to understand the heterogeneity of formations. Keep in mind that plotting such data in a biaxial graph to determine the nature of correlation is inherently wrong. When one variable shows an increase in values in a closed system, the remaining variable should automatically show a decrease in values. Therefore, we need to be cautious when using such data to infer relationships.

Unlike many other disciplines, directional variables are unique to geosciences, where they are critically important. We express such data in angles, such as the strike and dip of a geologic feature (i.e., fold and fault). These data require special methods of plotting and analysis because they have a circular distribution.

## 1.6.2 Non-Numerical Data Types

Apart from numerical data, we also use various data types in geosciences. Figure 1.8 shows other varieties of data used in ML, including image, text, audio, and video

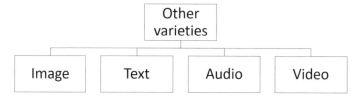

**Fig. 1.8** Other data types used in machine learning

data. Image data are very common in geosciences (for example, thin sections, CT scans, scanning electron microscopy, seismic, outcrop, and remote-sensing images). A picture comprises several pixels (2D) and voxels (3D) depending on its dimensions. Each pixel or voxel contains the fundamental properties of the image that can be analyzed statistically. Different open-source packages exist for processing and interpreting image data and may facilitate better statistical analysis of certain types of features, such as minerals, pores, and fractures. Deep learning is particularly useful in analyzing image data. As we move into more real-time analytics, audio and video data may become more mainstream.

In geosciences, we often deal with variables that are a combination of other variables of different types. For example, reservoir quality is an ordinal variable because it is based on a certain reference. It can be excellent, good, moderate, poor, or very poor. However, reservoir quality is also based on several other factors, such as grain size, porosity, and permeability, none of which are ordinal variables, but continuous ratio variables. In this context, we should also remember the common definition of reservoir quality as applied to conventional reservoirs is not always applicable to unconventional reservoirs (e.g., shale and tight sandstones/carbonates). Because unconventional reservoirs (including enhanced geothermal systems) typically require hydraulic stimulation, geomechanics, fracturability, organic matter content (for shale reservoirs), and heat flow are also considered when analyzing reservoir quality (see Mohaghegh 2017 for further details). Studies have shown that geomechanics plays a larger role in certain hybrid plays than many other parameters which are commonly recognized as important. In subsurface fluid storage studies (carbon, hydrogen, and wastewater), geomechanical and matrix properties are critical. The bottom line is that the definition of reservoir quality is relative, and the parameters it uses also vary from continuous to discrete nature.

During integrated geologic and reservoir modeling purposes, we must also be aware of the concepts of 'hard' data and 'soft' data. Hard data refers to field measurements (e.g., outcrops and subsurface). For example, mineralogy, fluid types, and volume are all hard data. Soft data corresponds to variables interpreted, estimated, or guessed by geoscientists and engineers. The knowledge and understanding of a specific depositional system (i.e., facies association rules) and hydraulic fracture geometry are examples of soft data. We often do not have access to hard data; in these cases, we can use soft data-based rules to make interpretations and generate geologically meaningful models.

## 1.7   Scales, Resolutions, and Integration of Common Geologic Data

In geosciences, we deal with multi-scale data. Data come from different scales of resolution, ranging from nanometers to thousands of kilometers. The concept of scale is fundamental in geologic data analysis. This concept is useful in every domain, especially in integrating data from multiple sub-disciplines, such as lab measurements, outcrop observations, geophysics, satellite measurements, etc. All these data are essential in understanding geology, depending on the project objectives. If we are interested in the regional picture of the study area, we can use coarse-resolution data, such as gravity, magnetic, 2D seismic, Lidar, and remote-sensing data. If we are interested in a particular location in a basin, we can use well logs, core, mud logs, and vertical seismic profiles from boreholes. If we are interested in high-resolution heterogeneity in rock samples, we can use petrographic thin sections, CT scans, micro-CT scans, and SEM images. We should keep in mind as the vertical resolution of a feature increases, its lateral resolution reduces. For example, seismic data has a good lateral resolution to identify regional-scale features, but it has a low vertical resolution compared to well logs and core data. Core and well logs have a high vertical resolution.

Here is an analogy: Think about a starry night. With the naked eye, we can gaze at the sky and count a few stars. Using a telescope, we can identify several of them correctly beacuse the telescope just enhanced our vision. Similarly, all the instruments used in subsurface geosciences can enhance our vision of geologic features at different scales, depending on what we want to see and how far they are from us, meaning how large or small they are.

The tools we typically use for geodata acquisition depend on whether we are interested in solving external heterogeneity or internal heterogeneity. In the industry, we are often interested in both, and therefore, proper data-driven integration is necessary. Although the word 'integration' is common, it is difficult to perform a true integration of multiscale geologic features. Broadly speaking, there are two types of integration efforts in geosciences, which I will elucidate with two distinct examples.

In the first approach, which is more common, geoscientists spend a lot of time describing and analyzing the geologic ground-truth, outcrop, core, cuttings, and thin section data. Then, when it comes to using that information in the well log or seismic domain, they simply attempt to observe the well log motifs or seismic reflector geometries and try to match them to the core- or outcrop-based information. This approach helps geoscientists understand the size and geometry of different geologic features visually. This kind of work is much needed at the exploratory phase or even to quality-check the ongoing modeling effort by intervening them with ground-truth data. Such methods are generally useful to conventional sandstone reservoirs. It is basically a visual pattern recognition exercise by the experts at a rudimentary level. As the experts change, so do the interpretations and their consistencies. Carbonate and mudstone reservoirs are complex, and conventional well logs cannot always capture the degree of diagenesis at a small scale.

The other approach, which is more recent and less common, is based on a 'fusion' approach. In this approach, we attempt to derive similar rock properties from different data sources, such as core, well log, and seismic. By deriving similar properties from physically independent sources, we do not bias the information from one data source to the other. Then, we perform upscaling/downscaling and geostatistics- and ML-based analyses to build consistent and quantitative subsurface models at different scales. Such an approach reduces inconsistencies in interpretation, quantifies uncertainties, and directs us to improve the model in areas that need attention. This could be done via further data collection or processing. Figure 1.9 shows an example.

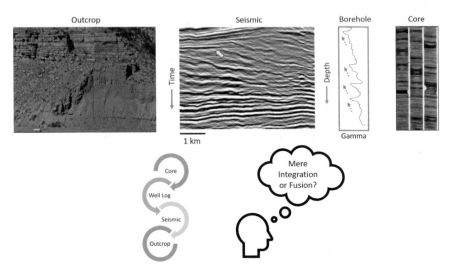

**Fig. 1.9** Conventional 'integration' versus a 'fusion' approach. The outcrop picture on the upper left is courtesy of Jonathan Rotzien

A fusion approach is critical to success in complicated mudstone reservoirs. Wang and Carr (2013) derived mineralogy and organic matter content from x-ray diffraction (XRD) and pyrolysis from core samples. They estimated the same mineralogy and total organic carbon (TOC) from advanced well logs and joint petrophysical inversion of conventional wireline logs. The joint inversion produced continuous multi-mineral solutions throughout the interval of interest, whereas core data were discrete and limited. Wang and Carr calibrated the core-based solution with log models and built integrated 3D lithofacies models at the basin scale using geostatistics and ML, with quantified uncertainties for each facies. Similarly, we can analyze fractures from micro-CT scans, CT scans, slabbed core samples, image logs, and perhaps very high-resolution 3D seismic data. These approaches are truly integrative and quantitative. Therefore, we must develop appropriate methodologies that will help in making efficient business decisions.

Ma (2019) correctly points out that we need both descriptive and quantitative geosciences working together. Perhaps we need to go beyond typical 'integration' and perform fusions of different datasets to build better subsurface models. By performing quantitative analysis, we can leverage well-established statistical concepts, test our hypotheses, and reduce inconsistencies and uncertainties in subsurface interpretations in a systematic and quantitative way. ML will play a big role in achieving these goals. We already utilize these methods with seismic inversion and prediction of rock properties (Hampson et al. 2001), in which we build a low-frequency geologic model based on seismic data and then perform certain statistical analyses to predict petrophysical properties (e.g., porosity) from inverted seismic data (i.e., impedance) away from the boreholes. This approach yields broad lateral coverage but at a higher resolution than the original seismic data. We will discuss some of these applications in Chapter 5.

That being said, well log motifs and seismic reflector patterns are still very useful elements, as deep learning shows massive potential to better analyze and capture images and shapes for further integration and analyses. We expect more research in this direction in the future. We should find a balance among new theories, conceptual models, and practicality. Because each theory and conceptual model has their own assumptions and limitations, and because real data come with all their assumptions and limitations, we need to be cautious in our data processing and integration efforts (Ma 2019). We can often achieve significant milestones with limited high-quality data in a laboratory that cannot be replicated in a massive field deployment scenario. Figure 1.10 shows the proportion of vertical versus lateral resolution of different types of data used in geosciences.

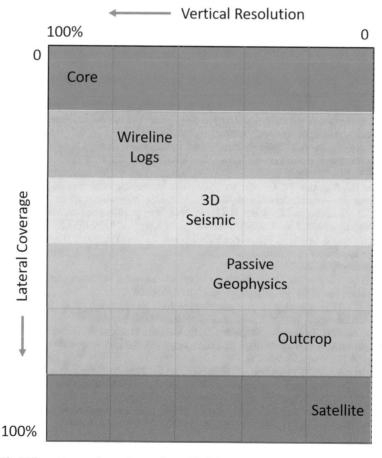

**Fig. 1.10** Different types of geosciences data with their proportion of vertical resolution versus their lateral coverage depicted

# References

Araya-Polo M, Dahlke T, Frogner C, Zhang C, Poggio T, Hohl D (2017) Automated fault detection without seismic processing. The Leading Edge 36(3):208–214. https://doi.org/10.1190/tle360302 08.1

Atkinson PM, Tatnall ARL (1997) Introduction neural networks in remote sensing. Int J Remote Sens 18(4):699–709. https://doi.org/10.1080/014311697218700

Bhattacharya S, Ghahfarokhi PK, Carr TR, Pantaleone S (2019) Application of predictive data analytics to model daily hydrocarbon production using petrophysical, geomechanical, fiber-optic, completions, and surface data: a case study from the Marcellus Shale, North America. J Petrol Sci Eng 176:702–715

Breiman L (2001) Random Forests. Mach Learn 45:5–32. https://doi.org/10.1023/A:101093340 4324

Carr TR (1982) Log-linear models, Markov chains and cyclic sedimentation. J Sediment Res 52(3):905–912. https://doi.org/10.1306/212F808A-2B24-11D7-8648000102C1865D

Conway D (2010) The data science venn diagram. https://drewconway.com/zia/2013/3/26/the-data-science-venn-diagram

Cortes C, Vapnik V (1995) Support-vector networks. Mach Learn 20:273–297. https://doi.org/10.1007/BF00994018

Davis JC (2002) Statistics and data analysis in geology. Wiley, New York

Doveton JH (1994) Geologic log analysis using computer methods. American association of petroleum geologists.

Gartner (2012) Information technology glossary. https://www.gartner.com/en/information-technology/glossary (accessed 2021)

Goodfellow IJ, Pouget-Abadie J, Mirza M, Xu B, Warde-Farley D, Ozair S, Courville A, Bengio Y (2014) Generative adversarial nets. Proceedings of the 27th international conference on neural information processing systems, Volume 2, pp 2672–2680

Govindarazu RS, Rao AR (2000) Artificial neural networks in hydrology. Springer

Hall B (2016) Facies classification using machine learning. Lead Edge 35(10):818–924. https://doi.org/10.1190/tle35100906.1

Hampson DP, Schuelke JS, Quirein JA (2001) Use of multiattribute transforms to predict log properties from seismic data. Geophysics 66(1):220–236. https://doi.org/10.1190/1.1444899

Hornik K, Stinchcombe M, White H (1989) Multilayer feedforward networks are universal approximators. Neural Netw 2(5):359–366. https://doi.org/10.1016/0893-6080(89)90020-8

Jaitly N, Nguyen P, Senior A, Vanhoucke V (2012) Application of pretrained deep neural networks to large vocabulary speech recognition. https://storage.googleapis.com/pub-tools-public-publication-data/pdf/38130.pdf

Krumbein WC, Dacey MF (1969) Markov chains and embedded Markov chains in geology. J Int Assoc Math Geol 1:79–96. https://doi.org/10.1007/BF02047072

Kuzma HA (2003) A support vector machine for AVO interpretation. SEG Technical Program Expanded Abstracts, 181–184. Society of Exploration Geophysicists.

LeCun Y, Boser B, Denker JS, Henderson D, Howard RE, Hubbard W, Jackel LD (1989) Back-propagation applied to handwritten zip code recognition. Neural Comput 1(4):541–551. https://doi.org/10.1162/neco.1989.1.4.541

Li H, Misra S (2017) Prediction of subsurface NMR T2 distributions in a shale petroleum system using variational autoencoder-based neural networks. IEEE Geosci Remote Sens Lett 14(12):2395–2397. https://doi.org/10.1109/LGRS.2017.2766130

Li J, Castagna J (2004) Support vector machine (SVM) pattern recognition to AVO classification. Geophys Res Lett 31(2):L02609. https://doi.org/10.1029/2003GL018299

Luthi SM, Bryant ID (1997) Well-log correlation using a back-propagation neural network. Math Geol 29:413–425. https://doi.org/10.1007/BF02769643

Ma YZ (2019) Quantitative geosciences: data analytics, geostatistics, reservoir characterization and modeling. Springer

McCormack MD (1991) Neural computing in geophysics. The Leading Edge 10(1):11–15. https://doi.org/10.1190/1.1436771

McCulloch WS, Pitts W (1943) A logical calculus of the ideas immanent in nervous activity. Bull Math Biophys 5:115–133. https://doi.org/10.1007/BF02478259

Minsky M, Papert S (1969) Perceptrons. Massachusetts Institute of Technology Press, Cambridge

Minsky M, Selfridge OG (1961) Learning in neural nets. Proceedings of the fourth London symposium on information theory (ed: Cherry C). Academic Press, New York, pp 335–347

Mitchell TM (1997) Machine learning. McGraw-Hill International

Mohaghegh SD (2017) Shale analytics: data-driven analytics in unconventional resources. Springer

Pal M, Mather PM (2003) An assessment of the effectiveness of decision tree methods for land cover classification. Remote Sens Environ 86(4):554–565. https://doi.org/10.1016/S0034-4257(03)00132-9

Qi L, Carr TR (2006) Neural network prediction of carbonate lithofacies from well logs, Big Bow and Sand Arroyo Creek fields, Southwest Kansas. Comput Geosci 32(7):947–964. https://doi.org/10.1016/j.cageo.2005.10.020

Rosenblatt F (1958) The perceptron: a probabilistic model for information storage and organization in the brain. Psychol Rev 65(6):386–408. https://doi.org/10.1037/h0042519

Ross ZE, Meier M-A, Hauksson E (2018) P wave arrival picking and first-motion polarity determination with deep learning. JGR Solid Earth 123(6):5120–5129. https://doi.org/10.1029/2017JB015251

Scher S (2018) Toward data-driven weather and climate forecasting: approximating a simple general circulation model with deep learning. Geophys Res Lett 45(22):12616–12622. https://doi.org/10.1029/2018GL080704

University of Toronto News (2012) Leading breakthroughs in speech recognition software at Microsoft, Google, IBM. https://www.utoronto.ca/news/leading-breakthroughs-speech-recognition-software-microsoft-google-ibm

Waldeland AU, Solberg AHSS (2017) Salt classification using deep learning. Conference proceedings, 79th EAGE conference and exhibition 2017, European Association of Geoscientists & Engineers, pp 1–5. https://doi.org/10.3997/2214-4609.201700918

Wang G, Carr TR (2013) Organic-rich Marcellus Shale lithofacies modeling and distribution pattern analysis in the Appalachian Basin. Am Asso Petrol Geol Bull 97(12):2173–2205. https://doi.org/10.1306/05141312135

Werbos PJ (1974) Beyond regression: new tools for prediction and analysis in the behavioral sciences, PhD dissertation, Harvard University, Cambridge, MA.

Wessel, P. (2007) Introduction to statistics and data analysis. https://www.soest.hawaii.edu/wessel/DA/index.html

Wu X, Liang L, Shi Y, Fomel S (2019) FaultSeg3D: using synthetic datasets to train an end-to-end convolutional neural network for 3D seismic fault segmentation. Geophysics 84(3):IM35–IM45. https://doi.org/10.1190/geo2018-0646.1

Zuo R, Xiong Y, Wang J, Carranaja EJM (2019) Deep learning and its application in geochemical mapping. Earth-Sci Rev 192:1–14. https://doi.org/10.1016/j.earscirev.2019.02.023

# Chapter 2
# A Brief Review of Statistical Measures

**Abstract**  As geoscientists, we use a variety of data types (see Chapter 1). To analyze different varieties of geodata, we must use different statistical measures. A thorough understanding of various statistical measures and their applications to data analysis methods is important. In this chapter, we will learn about fundamental concepts in statistics and different data analysis measures. This chapter begins with the basic concept of random variables, which are widely used in statistics. We then move on to univariate, bivariate, time series, spatial, and multivariate graphing and analysis techniques.

**Keywords**  Random variable · Univariate data analysis · Bivariate data analysis · Time series analysis · Markov chain · Spatial data analysis · Variogram · Multivariate data analysis

## 2.1  Random Variable

The concept of random variable is critical to understanding statistics and machine learning (ML). 'Random' is a term used frequently in statistics. Randomness is everywhere, from geologic processes (i.e., deposition and diagenesis) to their products (i.e., minerals and rocks). A random variable or variate is a quantity that may take any of the values within a given set. The values are "statistically" generated based on some unseen probabilistic mechanisms and natural ordering. Any geologic parameters can be considered as a random variable. For example, porosity of sedimentary rocks generally ranges between 0 and 40%. Numerous experiments on rocks in several sedimentary basins around the world have established this observation. Suppose we have a new rock sample and must determine its porosity. We can tell the probable range (the spread between the variable's minimum and maximum values) of porosity; however, we cannot give the exact value of that sample's porosity without performing a new experiment. Another example of a random variable is the proportion of calcite in limestone. Based on experiments, we know that this proportion can be any value between 0 and 100%, but we do not know the exact proportion in a sample prior to an

**Fig. 2.1** The concept of a
random variable

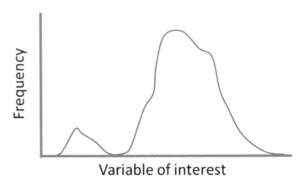

investigation of the sample. The same concept applies to all other related variables, such as permeability, fluid saturation, grain size, etc. (Fig. 2.1).

## 2.2   Common Types of Geologic Data Analysis

We can classify standard statistical and other data analytical techniques into at least five types based on the problem, the number of variables and types of variables being analyzed. These types include univariate, bivariate, time series, spatial, and multivariate analysis (Fig. 2.2). The measures of analysis are unique to each analysis. We use different statistical measures to describe the main characteristics of the dataset. We can use any statistical software or open-source programming (such as Python, R, and Julia) to plot the data and perform univariate analysis. In this book, I mostly use Python and Microsoft Excel™ for data plotting and analysis.

### 2.2.1   Univariate Analysis

Univariate data analysis is perhaps the most common type of analytical technique. In univariate analysis, we analyze each variable of interest individually. If the full

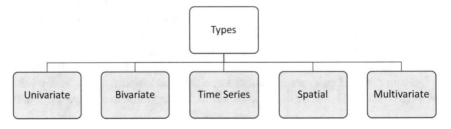

**Fig. 2.2** Common types of data analysis in geoscience

dataset contains several variables, we examine one variable at a time, independent of other variables. Variables we might analyze with univariate analysis include porosity, permeability, fluid saturation, acoustic impedance, length, and width of samples, etc.

Because our sampling of geologic materials, especially from the subsurface, is limited in number, size, and proportional distribution due to financial and logistic constraints, univariate measures provide us with a way to summarize the rudimentary statistical features of the individual variables in a dataset. We can use part of that knowledge (e.g., mean, median, and range) to predict a random variable's behavior in a new experiment or location. For example, suppose the porosity measurements from several sandstone samples range between 10 and 25%. In that case, we can use this information to predict the next sample's porosity if we do not have access to the core or even calibrate to the available geophysical logs.

There are several univariate measures for analyzing the statistical features of the variables. Most commonly, we use measures, such as mean, median, mode, standard deviation, and variance, to describe a dataset's characteristics. However, other measures, such as skewness and kurtosis, are also equally important because real-world data is not symmetric and does not follow a Gaussian distribution.

In univariate analysis, we can represent the data as a series of points or data along a line (Swan and Sandilands 1995). Several graphing techniques are used in univariate analysis, including histograms, bar diagrams, and box plots. A histogram is a graphical representation of frequency distribution. It is essentially a bar plot of a frequency distribution that is organized in intervals or classes. Histograms help us to characterize the average position, dispersion, and shape of a distribution.

In an ideal unimodal, symmetric distribution, the average position of the data points is near the center of the distribution (Fig. 2.3a). Therefore, the values of the mean, median, and mode are identical. However, distributions from subsurface data are not always symmetric in nature (Fig. 2.3b). Hence, the values of the mean, median, and mode are not the same. There are three different measures of mean: arithmetic, geometric, and harmonic. Arithmetic mean is the most common measure of the central location of the dataset. It is the sum of all data points divided by the number of observations. Arithmetic mean provides an unbiased estimate of the

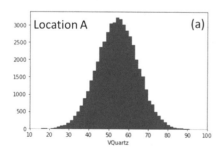
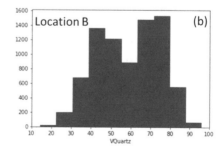

**Fig. 2.3** Histograms showing distribution quartz volumes at two different locations. **a** Unimodal and symmetric distribution and **b** asymmetric, bimodal distribution

population mean; however, it can be affected by outliers (equation below). Median is the midway value in a frequency distribution. Median is particularly useful in skewed distributions and datasets with outliers. Mode is the most frequent value of a distribution.

$$\overline{x} = \frac{\sum_{i=1}^{n} x_i}{n} \tag{2.1}$$

Where $\overline{x}$ is the arithmetic mean of n number of variable x. Consider, the porosity of several sandstone samples from a reservoir, in which $\phi_1$ (%) = [10, 10.2, 10.5, 10.9, 11.2, 11.6, 11.8, 12, 12.6, 12.9]. In this dataset, the mean is 11.33, and the mode and median are 11.2. In this case, the mean, mode, and median values are close to each other if not identical. We should keep in mind that mean is affected by outliers (positive and negative values), whereas the median is not. The median is unaffected because it lies at the midpoint of the full dataset, whereas the mean considers the minimum and maximum values of the dataset.

Consider the porosity of sandstone samples from the same reservoir with a few outliers added, $\phi_2$ (%) = [−999.25, −999.25, 10, 10.2, 10.5, 10.9, 11.2, 11.6, 11.8, 12, 12.6, 12.9, 999.25]. In this case, the median is still 11.2, but the mean is −68.5. The porosity of a rock cannot be negative; therefore, we need to remove outliers (i.e., -999.25). The same concept applies to very high values (i.e., 999.25). We will discuss this step further in the data processing section in Chapter 3. We can use box plots to show the presence of outliers in a dataset graphically.

Now, consider the porosity of vuggy carbonate samples from another reservoir, in which $\phi_3$ (%) = [10, 11, 11.5, 12, 12.4, 12.6, 26, 28, 29.5, 30]. In this case, the mean is 18.3, and the median is 12.5. This dataset is interesting as it reveals two different clusters of porosity values, the first ranging from 10 to 12.6% and the second from 26 to 30%. As geoscientists, we can easily interpret that the second cluster is probably more representative of the vugs in the carbonate samples. Why does this matter? First, the mean value, in this case, is not associated with any real values in the dataset. It lies at the dataset's central position, but this is far from both clusters. Such mean values are of no practical use. We could divide the full dataset into two subsets, corresponding to the two clusters (with vugs and without vugs), and then we could perform statistical computations for each subset individually. Second, there are quite a few porosity values higher than 26%. However, in this case, the values seem reasonable based on geologic knowledge and experience; these are not outliers. Therefore, a good data analytics model should build upon the data per cluster and the domain expertise.

We should keep in mind that the graphical appearance of a histogram depends highly on the bin size. If the bin size is large (i.e., coarse), the resulting histogram may not reveal the detailed data variation at different intervals. Although the results are reproducible with repeated sampling, such histograms are of limited value. In the case of a histogram with many bins (i.e., fine), the resulting histogram will reveal many details regarding the frequency distribution at various intervals, which is essential (Fig. 2.4).

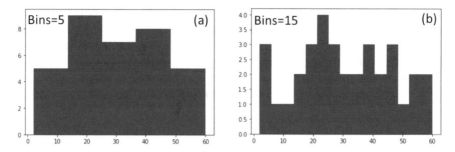

**Fig. 2.4** Histograms with different bin sizes of a porosity ($\phi_1$) dataset. Note how the different bin sizes change the histogram's appearance (**a** versus **b**)

Now, we will consider the scatter or spread of a dataset from the central position. An understanding of the data spread is important to characterize an experiments' preciseness and the robustness of the model that will be generated based on the data. If the scatter is high, more experiments may be needed to obtain a reliable average value of the variable. We use statistical measures, such as range, standard deviation, and variance to quantify data spread. Range refers to the difference between the maximum and minimum values of a variable. Consider the previous dataset of the porosity of sandstone samples, in which $\phi_1$ (%) = [10, 10.2, 10.5, 10.9, 11.2, 11.6, 11.8, 12, 12.6, 12.9]. In this case, the range is equal to the maximum value [12.9] minus the minimum value [10], which equals 2.9. A 2.9% variation in porosity is reasonable. However, if we consider the second dataset, $\phi_2$, the range is meaningless (Fig. 2.5). The presence of outliers (including null values) affects range and its effectiveness in statistical analysis.

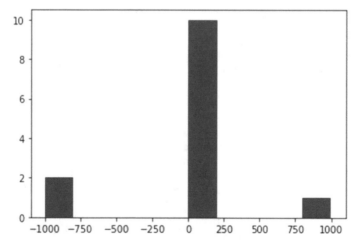

**Fig. 2.5** Histogram of a porosity ($\phi_2$) dataset with outliers

We commonly use standard deviation (SD) and variance ($\sigma$) to quantify data spread (equations below). Variance is the square of standard deviation. In essence, these two measures quantify the variability around the mean value of the variable. It is the difference between the mean of the squares and the square of the mean. If a dataset has a high standard deviation or variance, the values are spread out over a wide range, whereas if the dataset has a low standard deviation or variance, the values are close to the mean.

$$SD = \sqrt{\frac{\sum (x_i - \overline{x})^2}{n}} \; for \; population \; SD, \; use \; (n-1) \; instead \; of \; n \; for \; sample \; SD$$
(2.2)

$$\sigma = \frac{\sum (x_i - \overline{x})^2}{n} \; for \; population \; \sigma, \; use \; (n-1) \; instead \; of \; n \; for \; sample \; \sigma \quad (2.3)$$

In geophysics, we commonly use coherence attribute to delineate the boundaries between different geologic features (Chopra and Marfurt 2007). Coherence is one over variance. As opposed to variance, coherence measures the degree of similarity between adjacent samples. If coherence is high, samples are similar; if coherence is low, the samples are different. The maximum and minimum values of coherence are zero and one. Figure 2.5 shows the similarity of geologic features quantified using a coherence attribute on seismic data. It is always the best practice to quantify the standard deviation or variance, not just the average value, when describing a variable (Fig. 2.6).

Let us consider the business implications of standard deviation in a simplistic scenario. Suppose a formation is producing oil in two fields, A and B (Fig. 2.7). A company wants to decide where to invest and make confident decisions. The company is making its decisions based on the initial production (IP) rate. Figure 2.7 shows that the IP rate from the formation in field B is higher than that in field A, but

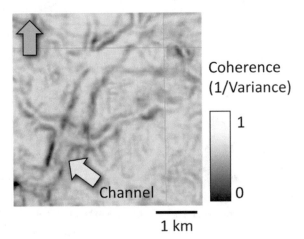

**Fig. 2.6** Coherence attribute on a stratal slice showing a channel system

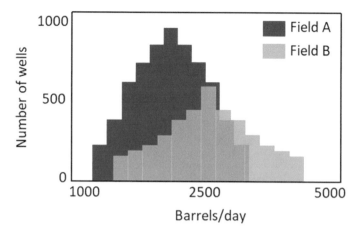

**Fig. 2.7** Distribution of daily initial production from a formation in two fields, A and B

there is a large standard deviation, which might be due to geologic and engineering parameters. However, the average IP from the formation is smaller in field A but with a small standard deviation. Although this is a multivariate problem, based on the IP rate only, the company should invest in the area with lower standard deviation because this will reduce the chance of failure.

Many times, we encounter datasets with the same central location and dispersion, but different shapes. Shape is a subtle property with implications in geologic data analysis. We use measures such as skewness and kurtosis to measure the asymmetry in data. Skewness is a measure of the asymmetry of the tails of a distribution. Distributions with positive skewness have large tails that extend toward the right (Fig. 2.8a). In contrast, distributions with a negative skew indicate that the distribution is spread out more to the left of the average value (Fig. 2.8b). For the first dataset, $\phi_1$, the skewness is 0.27, making it positively skewed. Shape analysis is also important in log-based sequence stratigraphy (i.e., fining-upward and coarsening-upward sequences).

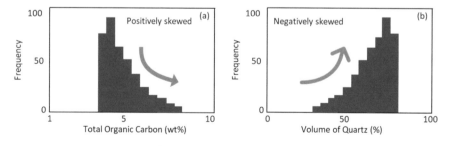

**Fig. 2.8** An example of **a** positively skewed distribution and **b** negatively skewed distribution

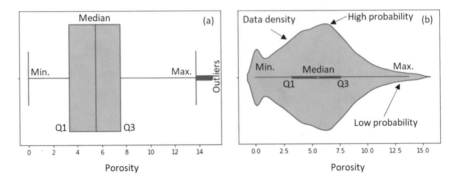

**Fig. 2.9** Porosity distribution in a mudrock, shown by **a** a box plot and **b** a violin plot. Mudrocks have lower porosity than sandstones and limestones

Apart from the commonly used histograms, other plots, such as box-and-whisker plots (or box plots) can represent more complete visualizations of univariate data. Box plots are a quick way to summarize data distribution (Fig. 2.9a). A box plot consists of five statistical measures: minimum and maximum values, the upper (Q1) and lower (Q3) quartiles, and the median. To create a box-and-whisker plot, we partition the whole data distribution into quartiles, mark the median value, and then draw lines ('whiskers') at both ends of the first and third quartile to reach the extreme ends (i.e., minimum and maximum values). The difference between the minimum and maximum values provides us with the range of the data distribution. Often, box plots are useful for detecting outliers present in a dataset. This makes them an important step in the exploratory data analysis and data pre-processing stage prior to deploying ML algorithms. If outliers are not detected and removed from the data, the ML model results will not be useful in terms of prediction and prescription.

Since the boom of ML applications in the physical sciences and widespread use of open-source programming languages such as Python and R, several other types of graphs have emerged to represent the data better and tell a more compelling story. The violin plot is an extension of the common box plot. In addition to the box plot's statistical measures, the violin plot also provides data density estimates (Fig. 2.9b). It is important to understand the shape of the data in a violin plot. Wider sections of a violin plot indicate a higher probability of that value, and thinner portions indicate lower probability.

### 2.2.2   Bivariate Analysis

In bivariate analysis, we analyze two variables together to understand their interrelations. The variables do not need to depend on each other and can be independent. However, they must have been acquired from the same system or object, otherwise, the results will have no meaningful implications. Bivariate analyses are widespread in subsurface characterization and modeling, for example, the relationships between porosity and permeability, acoustic impedance and porosity, porosity and water saturation, etc.

There are several standard bivariate measures to analyze the relationship between two variables: covariance (Cov), Pearson's correlation coefficient ($R_P$), and Spearman's rank correlation coefficient ($R_S$). Equations below show the mathematical expressions of these measures.

$$Cov(x, y) = \frac{\sum(x_i - \bar{x})(y_i - \bar{y})}{n} \; for \; population \; Cov, \; use \; (n-1) \; instead$$
$$of \; n \; for \; sample \; Cov \qquad (2.4)$$

$$R_P = \frac{cov(x, y)}{\sigma_x \sigma_y} = \frac{\sum(x_i - \bar{x})(y_i - \bar{y})}{\sqrt{\sum(x_i - \bar{x})^2 \sum(y_i - \bar{y})^2}} \qquad (2.5)$$

$$R_s = 1 - \frac{6 \sum_{i=1}^{n} d_i^2}{n(n^2 - 1)} \qquad (2.6)$$

Where x and y are two variables, n is the number of samples, σ represents variance, and d represents the distance between the ranks of x and y variables. Covariance indicates the joint variation of the two variables under study. It is the summed product of the deviation of two variables from their mean divided by the number of samples. Unlike variance, covariance can be either positive or negative.

Correlation coefficients provide us with a rough estimate of the relationship in a bivariate dataset. Depending on the dataset, the relation between two variables can be positive, negative, linear, nonlinear, or even a combination of these in certain points of the dataset. In simple cases, correlation coefficient is a useful metric for identifying the relationship between two variables and looking for causal mechanisms to explain that relationship. However, keep in mind that correlation is not causation.

Pearson's correlation coefficient ($R_P$) is a measure of the degree of linear correlation (Fig. 2.10). To compute Pearson's correlation coefficient, we must find the

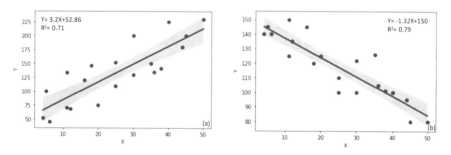

**Fig. 2.10** Bivariate relationships between two variables (X and Y) in different datasets: **a** the positive relationship between X and Y and **b** the negative relationship between them

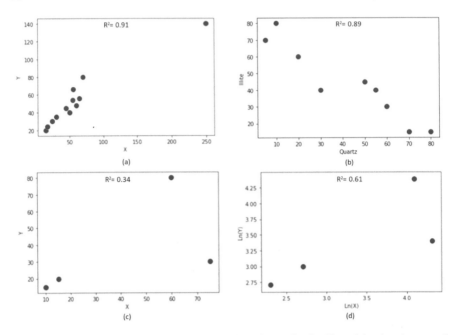

**Fig. 2.11** Examples of spurious correlations: **a** effect of an outlier, **b** effect of the closed nature of the dataset, and **c** and **d** effect of data transformation on correlation coefficients

number of samples, the mean deviation, and each variable's standard deviation. This coefficient is dimensionless, and its value can range between −1 and 1.

We must keep in mind that the correlation coefficient measures are subject to several controls, such as outliers, data transformation, and the closed nature of the data. The presence of extreme outliers affects the correlation coefficient significantly and thus can influence the model's predictive performance. Figure 2.11a depicts an example of such cases. To achieve accurate statistical inferences, we must detect outliers and remove them from the dataset.

Another common case of flawed correlation arises when we work on closed data, in which the sum of all the variables either equals 1 or 100%. For example, we would like to understand the relation between quartz and clay in sandstones. A scatter plot of these two variables will show a negative relationship because when one variable increases, the other should automatically decrease. Therefore, the premise of such a correlation is flawed. Figure 2.11b shows an example.

Data transformation can be a necessary step in ML-based analysis. However, we should also keep in mind that certain operations or transformations of variables (such as natural logarithms) can affect correlation. In these cases, the correlations between variables do not reflect the true relationship between those variables. Figure 2.11c and d show such examples.

Another important aspect of using correlation coefficient-based information for modeling is related to the boundary conditions under which the variables were

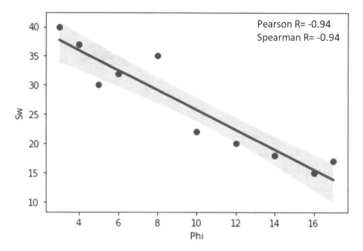

**Fig. 2.12** Regression plot of the data (porosity [Phi] and water saturation [Sw]) from Table 2.1, with corresponding Pearson's and Spearman's correlation coefficient values

acquired. These conditions have major implications while deploying ML algorithms for prediction. Suppose we are working on predicting fluid flow from a reservoir. Fluid flow changes from transient to boundary-dominated flow over time. If we have data from one of these two domains, the results will not be applicable to the other domain. We must be informed about the conditions and domains in which the variables were collected prior to data analysis to make well-founded data-driven decisions.

Sometimes, we are interested in understanding the relationship between the ranks of variables under study. In such cases, we use Spearman's rank correlation coefficient ($R_S$) instead of Pearson's correlation coefficient. $R_S$ is a measure of statistical dependence between the ranks of two variables. We compute this coefficient by calculating the correlation coefficient between the ranks of the original variables. We can use $R_S$ for applications such as understanding the relationship between the ranks of porosity and permeability to characterize reservoir quality (Fig. 2.12 and Table 2.1).

## 2.2.3  Time Series Analysis

Time series is similar to bivariate data, except that the variable on the y-axis varies with respect to time on the x-axis. The main variable of interest changes with time. An ideal time series has a few basic features, including seasonality, stationarity, and autocorrelation. Seasonality corresponds to periodic fluctuations in the data, for example, household gas demand in the winter versus summer. Stationarity means the statistical properties of the variable (e.g., mean and variance) do not change over time.

**Table 2.1** Porosity and water saturation data for computing Spearman's rank correlation coefficient

| Phi | Sw | R(Phi) | R(Sw) | $d_i$ | $d_i^2$ |
|---|---|---|---|---|---|
| | | | | R(Phi)−R(Sw) | [R(Phi)−R(Sw)]² |
| 16 | 15 | 9 | 1 | 8 | 64 |
| 17 | 17 | 10 | 2 | 8 | 64 |
| 14 | 18 | 8 | 3 | 5 | 25 |
| 12 | 20 | 7 | 4 | 3 | 9 |
| 10 | 22 | 6 | 5 | 1 | 1 |
| 5 | 30 | 3 | 6 | −3 | 9 |
| 6 | 32 | 4 | 7 | −3 | 9 |
| 8 | 35 | 5 | 8 | −3 | 9 |
| 4 | 37 | 2 | 9 | −7 | 49 |
| 3 | 40 | 1 | 10 | −9 | 81 |
| | | | | Total | 320 |
| | | Spearman's rank correlation coefficient: −0.94 | | | |

An example of stationarity in geology is varve deposition in lakes. Autocorrelation helps us identify the presence of cyclicity in a time series.

There are some fundamental concepts in time series analysis that are highly applicable to geosciences. Using time series, we can find various types of interesting relationships in data, such as cyclic, acyclic, exponential, and power relationships. As geoscientists, we are interested in whether there are patterns or trends in data, whether we can quantify any periodicities, and whether we can use the information for forecasting the behavior of a variable. Geoscientists use time series in many areas, including sea-level change, daily temperature changes, fluid flow, and paleoclimate analysis (Fig. 2.13). With the advent of multi-lateral well drilling and multi-stage borehole completion technologies in resources exploration and development, it has become more critical than ever to analyze the efficacy of hydraulic stimulation and

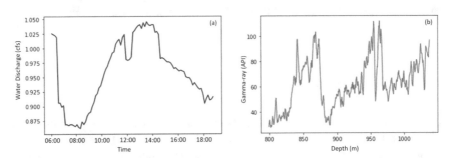

**Fig. 2.13** Examples of time series, **a** water discharge over time and **b** gamma-ray log over depth (one may also call it a depth series)

determine which geologic zones can produce more hydrocarbons. Streaming micro-seismic and fiber-optic (DTS and DAS) data are prime examples of areas that make use of time series analysis and require future research (Amini et al. 2017; Ghah-farokhi et al. 2018). This is also critical to long-term carbon and hydrogen storage programs in subsurface.

There are several important factors we need to keep in mind while analyzing time series data or applying ML to analyze such signals, many of them critical to geosciences. Time series analysis requires data acquired at equal intervals, and the data should have a high signal-to-noise ratio. It is not always possible to collect good data at equal intervals in the subsurface and on outcrops due to logistics, cost, time, access, and instrumental issues. In such cases, we can use several signal-processing techniques, such as interpolation, smoothing, and filtering, to remove artifacts. In another scenario, we might have a combination of variable sedimenta-tion rate, including a hiatus (e.g., unconformity) in an area. In such cases, the infer-ences drawn from time series analysis would not be insightful. The resolution of the instrument being used to record time series data is also important to consider. Take an example of the recording of micro-earthquakes (magnitude < 2.0). In general, commonly used seismographs do not record micro-earthquakes, but studies show that micro-earthquakes are important in analyzing induced seismicity because the accumulation of several tremors over time can lead to a big earthquake. Therefore, recording and analysis of micro-earthquakes are useful in disaster mitigation and operational management.

### 2.2.3.1 Markov Chains

Markov chains are a useful concept in time series analysis. A Markov chain describes a sequence of possible events in which each event's probability depends on the state attained by previous events. This unique feature enables us to analyze the nature of transitions from one state to another in a variable of interest.

Many geologic data can be considered as a succession of different states over time (or depth), such as facies variation. Markov chains are very useful for analyzing stratigraphic successions in which we might expect cyclic patterns (Fig. 2.14). We expect these patterns in a variety of depositional settings, including deltaic and lacus-trine deposits. The Markov chain's utility lies in analyzing the transition frequency and probability, which can be further used in predicting the pattern of change over time and predicting samples that might be missing from a section. The transition frequency matrix in Markov chain expresses the number of transitions from one state to another (including self-transitions), and the transition probability matrix quantifies the tendency for one state to succeed another at a fixed sample rate (Wessel 2007). We compute transition probability by dividing each row of transition frequency by its row total. Figure 2.14 shows a line plot of different well logs, defined by facies. Table 2.2 shows an example of transition frequency and transition probability using the data from Fig. 2.14.

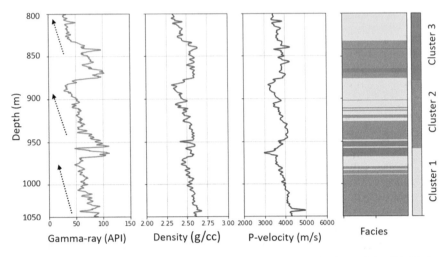

**Fig. 2.14**  A plot of three well logs (gamma-ray, density, and P-wave velocity) and simplified facies in a shaly sand formation. Yellow color in the fourth track corresponds to sandy facies, whereas gray indicates more shaly facies. Black dash arrows indicate coarsening-upward (or cleaning-upward) sequence. The log plot shows the presence of cyclicity of log curves and facies. Markov chains can be used to quantify cyclicity

**Table 2.2**  An example of Markov chain transition frequency and transition probability using data from Fig. 2.14. It shows a high probability of self-transition of cluster 1 (177 times with a 0.94 transition probability) and a relatively low probability of transition for cluster 1 to cluster 2 (nine times with a 0.05 transition probability), compared to cluster 1 to cluster 3 (two times with a 0.01 transition probability)

| | Cluster 1 | Cluster 2 | Cluster 3 |
|---|---|---|---|
| Cluster 1 | 177 | 9 | 2 |
| Cluster 2 | 10 | 233 | 2 |
| Cluster 3 | 0 | 4 | 40 |

Transition Frequency

| | Cluster 1 | Cluster 2 | Cluster 3 |
|---|---|---|---|
| Cluster 1 | 0.94 | 0.05 | 0.01 |
| Cluster 2 | 0.04 | 0.95 | 0.01 |
| Cluster 3 | 0 | 0.09 | 0.9 |

Transition Probability

### 2.2.3.2   Autocorrelation

We commonly use autocorrelation, cross-correlation, and spectral analysis for time series analysis, for example, signal processing. Autocorrelation is the correlation between a time series and its duplicate at different time lags (or shifts). It is a standard technique for analyzing cyclicity in individual variables. By computing autocorrelation over all possible lags, we can construct an autocovariance function that varies with time. This function is useful to detect self-similarity. Figure 2.15 represents autocorrelograms corresponding to the datasets in Fig. 2.13.

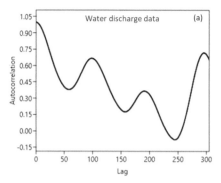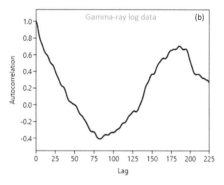

**Fig. 2.15** Autocorrelograms for the data used in Fig. 2.13. Curve patterns indicate cyclicity present in the datasets

### 2.2.3.3 Cross-Correlation

Whereas autocorrelation is useful in analyzing self-similarity, cross-correlation is useful in analyzing the relationship between two time series datasets. In essence, cross-correlation measures the degree of similarity between two signals while shifting one of them in time (over lag). This technique helps us identify hidden relationships between two time series. With cross-correlation, it is possible to analyze the causal relations between two time series. Davis (2002) gives an excellent example of the correlation between waste injection and the occurrence of earthquakes over time in the Rocky Mountain Arsenal. Injection of waste water into the subsurface increases pore pressure, which reduces the effective stress that can cause fault slip and gives rise to induced seismicity. We can use the time lags derived from such cross-correlations to make decisions in the field to mitigate the potential of injection-induced seismicity. However, the utmost care and domain expertise are needed before making inferences about underlying causal mechanisms (Fig. 2.16).

## 2.2.4 Spatial Analysis

Spatial data are perhaps the most common data that geoscientists deal with on a regular basis. One can say that we, geoscientists, are paid to make maps and interpret them. Therefore, maps should be meaningful, consistent, predictive, and useful for reaching actionable decisions. If a map does not have these features, the map should be discarded.

Dealing with spatial data requires expertise in the relevant domain and computational literacy. In general, spatial data is composed of three parameters: latitude (x), longitude (y), and the value of the variable of interest (z), which we can use to generate maps (Fig. 2.17). In the case of subsurface data, we use latitude, longitude, depth, and the value of the variable to generate property maps and 3D models.

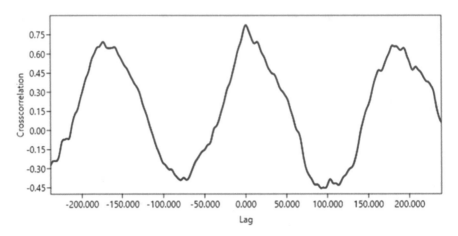

**Fig. 2.16** Cross-correlogram between gamma-ray and density logs for the dataset in Fig. 2.14. Note the cyclicity present in these two curves, which is due to the repetitive pattern of facies

Understanding the nature of the variable and its spatial correlation limits is necessary for generating maps, regardless of hand-contoured and computer-generated maps. Although we do not need the specific value of the variable's spatial correlation limit in hand contouring, we do use it based on the patterns of the observed data and gaps, our experience, and an understanding of the probable geologic model. For computer-generated maps, we need to determine the spatial correlation limit to make meaningful and predictive maps. The variogram (or semivariogram) provides us with the spatial continuity or roughness of a dataset (Deutsch and Journel 1992).

Variograms are at the heart of geostatistics. Variogram analysis consists of an experimental variogram calculated from the data and the variogram model fitted to the data. In essence, it is the variance between samples at a specified interval or distance apart along different directions. A variogram consists of three elements: range, sill, and nugget. Depending on the problem and data availability, we can construct variograms in different directions, horizontally and vertically. A variogram can be mathematically expressed as.

$$\gamma_h = \frac{\Sigma[z(u) - z(u + h)]^2}{2N(h)} \tag{2.7}$$

In which $\gamma$ represents the semivariogram, $h$ refers to the lag distance, $u$ indicates the data locations, and $N(h)$ corresponds to the number of pairs for lag $h$.

Sill represents the value at which the model first flattens out. Range corresponds to the distance at which the model first flattens out. In other words, the range is the maximum distance at which data are correlated. Nugget is the value at which the semivariogram (almost) intercepts the y-axis. The nugget represents variability at distances smaller than the typical sample spacing, including measurement error. We need all these metrics to define a variogram and create computer-generated maps

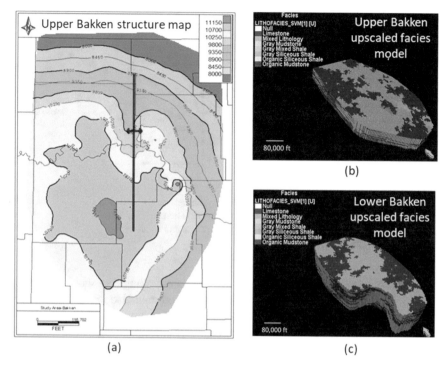

**Fig. 2.17  a** Structure map of the upper Bakken member in the United States. **b** and **c** show the 3D mudstone facies models (after Bhattacharya and Carr 2019). Sequential indicator simulation was used to generate 3D facies models of the upper and lower Bakken. Both models (**b** and **c**) are flattened on the top using a reference horizon to better visualize facies variation (Reprinted from Journal of Petroleum Science and Engineering, 177, S. Bhattacharya and T.R. Carr, Integrated data-driven 3D shale lithofacies modeling of the Bakken Formation in the Williston basin, North Dakota, United States, 1072–1086, Copyright (2019), with permission from Elsevier)

and models. Figure 2.18 shows different variograms with different sill, range, and nugget values.

There are several mathematical models of variogram, including linear, spherical, exponential, and Gaussian, etc. (Fig. 2.18b). Each of these models behaves differently and has different output maps. Although spherical and exponential variograms have similar behavior near the origin, exponential variograms climb faster than spherical variograms. Gaussian variograms yield very smooth results. An exponential variogram may show a very high degree of heterogeneity, which may be useful for rock properties.

Gorsich and Genton (2000) suggested computing derivatives and using them to select what kind of variogram models should be used. We must ensure the chosen mathematical model fits the observed data. Based on the apparent fit, we must determine the sill, range, and nugget values, which we can use for mapping and modeling. Different variogram models applied to the same data will result in different maps,

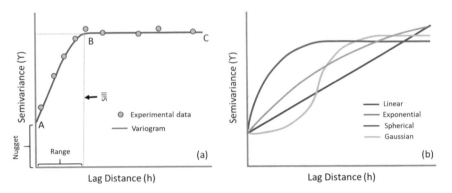

**Fig. 2.18** **a** An experimental semivariogram; the region between A and B is spatially correlated, whereas the region between B and C is not correlated. The shape of the curve can be used to determine range, sill, and nugget values. **b** Various types of empirical variogram models which can be used based on the data pattern.

which will have significant implications for decision-making. We should also keep in mind that fitting mathematical models to observed data also requires subjective judgment and previous experience. Nonetheless, variograms should be data-driven (using good data) and based on geologic information.

### 2.2.5  Multivariate Analysis

Multivariate analysis applies to more than three variables. More variables mean more dimensions. High dimensionality means the dataset has many features. As mentioned in Chapter 1, geosciences—specifically subsurface geosciences—have been experiencing a massive boom in data due to the advent and combined application of new drilling, completion, and sensor technologies. Additionally, geoscientists are also figuring out ways to mine data from a trove of old resources. Data can take various formats (numerical, image, text, audio, and video) or even a combination of formats. In such cases, commonly used measures in univariate and bivariate statistics are not very useful for visualizing and gleaning important information from data to make effective decisions. Conventional spreadsheet-based graphs (e.g., Excel™) are not helpful in analyzing multidimensional data. Although some authors treat multivariate data as an extension of bivariate data, this is not accurate.

Effective visualization is key to understanding a multivariate dataset. We can deploy different strategies to visualize multivariate data. These could be in the form of a pair plot (scatter matrix plot) or a plot in a reduced number of dimensions through principal component analysis. We can generate pair plots by combining scatter plots of all variables in a dataset. Figure 2.19 shows wireline log data from a mudstone formation in North America and the corresponding pair plot (Fig. 2.20). We can also add the correlation coefficients between each pair of variables, trend lines, or color

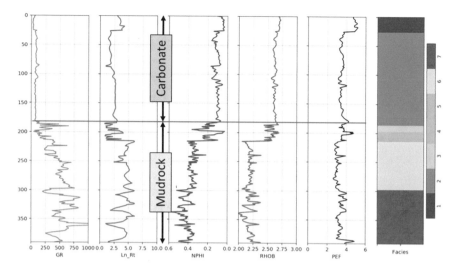

**Fig. 2.19** Plot showing conventional wireline logs and classified facies in a mudstone-carbonate succession. Y-axis corresponds to sample numbers (not the exact depth) with respect to an increasing order of facies (first track from right)

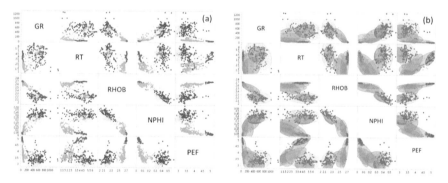

**Fig. 2.20** Pairplot or scatter matrix plot of conventional wireline log responses **a** with clusters and **b** without clusters

codes to indicate the relationships between different variables. Such plots provide us with an overview of the relationships among several variables and their patterns simultaneously.

Interestingly, such pair plots reveal that some of the clusters are overlapping, which cannot be resolved with such plots. Therefore, we must move beyond the regular measures used in conventional statistics (or STAT 101). This is where we begin using ML for data classification and prediction.

Principal component analysis (PCA) provides a convenient mechanism for visualizing multivariate data. It is a technique for reducing data dimensionality (Fig. 2.21). Humans cannot perceive multidimensional data, but we can reduce the number

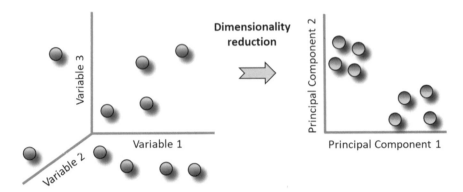

**Fig. 2.21** The simplified concept of principal component analysis for dimensionality reduction. Principal components are essentially scaled eigen vectors

of dimensions by combining the most important ones and removing the rest with PCA, allowing us to observe and analyze the essential features of the dataset. Principal components define a variance maximizing the mutually orthogonal coordinate system. Although there are several principal components, the first few principal components are adequate for describing most of the variability in data.

Suppose we are working with $n$ number of parameters for facies classification (e.g., seismic attributes, petrophysical properties, and geochemical data). If we graph such data, they will form a data cloud in the $n$-dimensional space. However, we cannot use this information with conventional techniques. PCA proceeds by first finding the axis along which the data is most spread out. It does this by computing eigenvectors. The first principal component (or the first eigenvector) accounts for the maximum amount of variability (or information), whereas the remaining components (or other eigenvectors) represent the rest of the information. We can then plot only those principal components which account for most of the information and analyze them in the scatter plot (Fig. 2.22). Thus, the dimensionality of an $n$-dimensional space shrinks to a lesser dimensional space, or principal component space, which still contains most of the information in the original dataset. Essentially, the principal components provide a new reference frame for looking at the data. By providing the most important principal components, PCA helps us in feature selection, outlier detection, and clustering.

Geoscientists have widely used PCA over the years. Qi and Carr (2006) demonstrate the application of PCA to lithofacies classification in a carbonate formation in Kansas. PCA analysis is used for unsupervised pattern recognition and discrimination. Zhao et al. (2015) show an example of PCA for seismic-attribute-based facies classification (Fig. 2.23).

We must keep in mind that some information may be lost after PCA. The examples could be rare facies within the dataset. Although the number of samples representing a facies may be small, this data is necessary for geologic mapping to truly understand the processes at work. If these rare facies do not influence the problem, we may

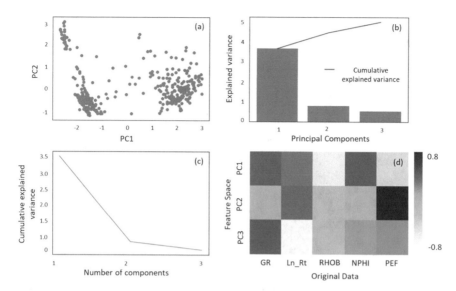

**Fig. 2.22** Principal component analysis plot corresponding to data from Fig. 2.19. **a** Scatterplot between principal components 1 and 2, **b** explained variance, **c** cumulative explained avarice (or scree plot), and **d** heatmap showing the relationship between original log data and feature space in PCA (**d**)

discard this information. Another common problem with PCA is its oversimplified analysis which reduces the dimensionality of the original data. Interpretation of PCA results can be very difficult because original variables are no longer used. Moreover, PCA works on the principle of linear relationships between variables, which is not relevant to numerous geologic problems. Therefore, we need to understand the scope of the problem, the nature of the data, and their representability to solve the problem before resorting to this technique.

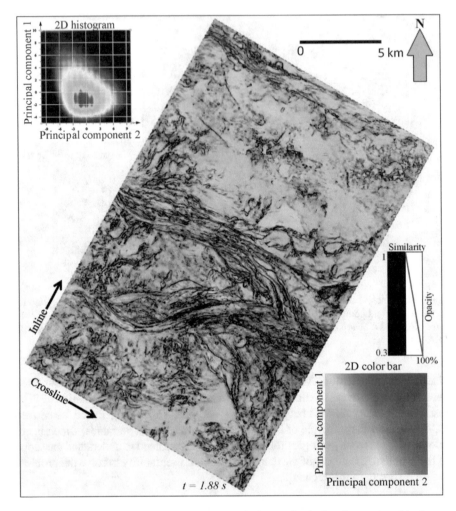

**Fig. 2.23** An example of principal component analysis on seismic data for stratigraphic feature identification (e.g., channels and point bars) in New Zealand. The 2D color bar corresponds to two principal components plotted here (Zhao et al. 2015) (Permission granted from SEG)

# References

Amini S, Kavousi P, Carr TR (2017) Application of fiber-optic temperature data analysis in hydraulic fracturing evaluation: a case study in Marcellus Shale. Unconventional resources technology conference, Austin, TX, 24–26 July 2017. https://doi.org/10.15530/urtec-2017-2686732

Bhattacharya S, Carr TR (2019) Integrated data-driven 3D shale lithofacies modeling of the Bakken formation in the Williston basin, North Dakota, United States. J Petrol Sci Eng 177:1072–1086. https://doi.org/10.1016/j.petrol.2019.02.036

Chopra S, Marfurt KJ (2007) Seismic attributes for prospect identification and reservoir characterization. Society of Exploration Geophysicists

Davis JC (2002) Statistics and data analysis in geology. Wiley, New York

Deutsch CV, Journel AJ (1992) GSLIB: geostatistical software library and user's guide. Oxford University Press, New York

Ghahfarokhi PK, Carr TR, Bhattacharya S, Elliot J, Shahkarami A, Martin K (2018) A fiber-optic assisted multilayer perceptron reservoir production modeling: a machine learning approach in prediction of gas production from the Marcellus shaleShale. Unconventional Resources Technology Conference, Houston, Texas, 23–25 July 2018, https://doi.org/10.15530/urtec-2018-290 2641

Gorsich D, Genton M (2000) Variogram model selection via nonparametric derivative estimation. Math Geol 32:249–270. https://doi.org/10.1023/A:1007563809463

Krumbein WC, Dacey MF (1969) Markov chains and embedded Markov chains in geology. J Int Assoc Math Geol 1:79–96. https://doi.org/10.1007/BF02047072

Qi L, Carr TR (2006) Neural network prediction of carbonate lithofacies from well logs, Big Bow and Sand Arroyo Creek fields, Southwest Kansas. Comput Geosci 32(7):947–964. https://doi.org/10.1016/j.cageo.2005.10.020

Swan ARH, Sandilands M (1995) Introduction to geological data analysis. Blackwell Science, Oxford

Wessel P (2007) Introduction to statistics and data analysis.

Zhao T, Jayaram V, Roy A, Marfurt KJ (2015) A comparison of classification techniques for seismic facies recognition. Interpretation 3(4):SAE29–SAE58. https://doi.org/10.1190/INT-2015-0044.1

# Chapter 3
# Basic Steps in Machine Learning-Based Modeling

**Abstract** Data-driven machine learning (ML) approaches are becoming very popular in analyzing facies, fractures, faults, rock properties, and fluid flow in subsurface characterization and modeling. Our reservoirs are becoming more data-rich due to the advent of new drilling, completion, and sensor technologies. Modeling of these variables from such data-rich reservoirs is a complex multivariate, multiscale, and multidisciplinary problem that we can handle with ML algorithms. In this chapter, we will learn about the fundamental steps in deploying ML models to solve our problems. Although there are several ML models available, including both traditional algorithms and deep learning algorithms, some steps are very similar. The selection of a particular algorithm over other depends on the data and the problem itself, complexity, interpretability, time, and cost. In this chapter, we focus on the nature of the problems and provide a systematic guide to building, evaluating, and explaining these data-driven models, irrespective of the algorithms.

**Keywords** Outlier detection · Unsupervised learning · Supervised learning · Semi-supervised learning · Classification problems · Regression problems · Modeling steps · Model performance · Model interpretation

## 3.1 Identification of the Problem

Identifying the problem is of paramount importance in ML-based modeling at the onset of a project. It will guide the selection, aggregation, abstraction, and transformation of sensitive input variables for robust output modeling that could give insights into complex processes. We can use the knowledge derived from such exercises for predictive and prescriptive purposes. Refer to Chapter 1 for the various types of tasks we can perform with ML. Based on the literature and our general scientific knowledge, we could effectively apply ML models to some of the subsurface geoscience problems as below:

1. Identifying borehole washout zones and their effects on geophysical log responses (*outlier detection problem*)

© The Author(s), under exclusive license to Springer Nature Switzerland AG 2021     45
S. Bhattacharya, *A Primer on Machine Learning in Subsurface Geosciences*,
SpringerBriefs in Petroleum Geoscience & Engineering,
https://doi.org/10.1007/978-3-030-71768-1_3

2.  Identifying the groups in seismic data for an exploration area with no ground-truth or calibratable data (*clustering problem*)
3.  Characterizing the vertical and lateral heterogeneities of a sedimentary formation to better understand the depositional and digenetic processes and basin-fill history (*classification problem*)
4.  Characterizing the presence of vugs in a carbonate reservoir to understand the degree of fluid-rock interactions and diagenetic processes for better prediction of sweet spots for resource exploration or fluid injection (*classification problem*)
5.  Analyzing the distribution of multiphase faults and fractures in an area and associating them with plate tectonics and stress directions over geologic time to better understand their implications on reservoir compartmentalization and deliverability (*classification problem*)
6.  Predicting the vertical and lateral variations in reservoir and geomechanical properties at the seismic scale for decision-making in an unconventional reservoir (*regression problem*)
7.  Predicting the occurrence of hydraulic frac-hits (yes or no) in an active resource development area due to variations in rock properties, drilling, and completion designs (*classification problem*)
8.  Analyzing the efficacy of hydraulic fracturing on fluid flow from individual stages of horizontal wells to better understand the controls on foot-scale (or meter-scale) geologic variations and completion designs (*regression problem*)
9.  Predicting the missing rock physics properties of a reservoir to guide the seismic inversion process for mapping variations in acoustic impedance and facies properties to delineate sweet spots (*regression problem*)

Classification problems are problems in which the response variable is discrete or categorical in nature. Clustering problems are classification problems in which we do not have access to ground-truth data to validate the models. The predictor variables can be either continuous or discrete; for example, the presence or absence of facies, fractures, faults, vugs, salt, and mass-transport deposit, etc. These problems are similar to Boolean algebra, in which we can assign a value of one if a condition is true and zero if false. These problems are of particular interest to geologists, geochemists, petrophysicists, and seismic interpreters. By applying ML to these models, we can quickly generate a map or a 3D model showing the distribution of an output variable. This information can be primarily used for understanding geologic processes and making decisions on drilling, resource recovery, and fluid storage. Clustering problems are useful at the exploration and appraisal stages when we have seismic and limited borehole data. We can use these datasets to better understand and perhaps refine the conceptual geologic model prior to field development.

Apart from traditional ML algorithms, deep learning algorithms are becoming more commonly applied to classification problems when the problems involve large data and images, such as seismic, advanced petrophysical logs (i.e., image logs and nuclear magnetic resonance [NMR]), and core photographs. We will discuss deep learning in Chapter 4. In the last few years, researchers have shown a variety of case studies on ML-based facies and fracture classification using well log and core data (Al-Anazi and Gates 2010; Wang and Carr 2012a, b; Bhattacharya et al. 2016; Howat et al. 2016; Li and Misra 2017; Bhattacharya and Mishra 2018). There has been a major uptick in deep-learning-assisted seismic interpretation of structure and stratigraphy (Huang et al. 2017; Alfarraj and AlRegib 2018; Di et al. 2018, Dramsch and Lüthje 2018; Zhao 2018; Alaudah et al. 2019; Di et al. 2019; Wu et al. 2019). Pires de Lima et al. (2020) showed an excellent example of fossil identification using deep learning. Figure 3.1 shows an example of a ML-based classification problem.

In the case of regression modeling, the response variable is continuous. The predictor variables (or input) can be either continuous or discrete. In geosciences, we can consider the output from regression models more like a sequence or a series in time, frequency, or depth domain; for example, reservoir properties (i.e., porosity, permeability, fluid saturation), geomechanical properties (i.e., Young's modulus, Poisson's ratio, etc.), and hydrocarbon production, etc. These problems are useful

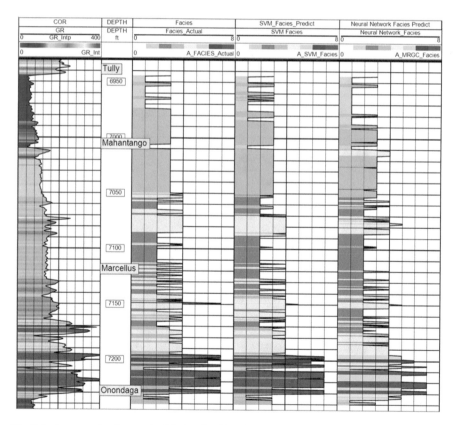

**Fig. 3.1** An example of supervised facies classification results from support vector machine, SVM (third track) and artificial neural network, ANN (fourth track), compared against the core-log defined facies (second track) in a well in the Appalachian Basin, North America. The SVM-based facies match the original facies better than the ANN-based facies

to geoscientists and engineers interested in predicting missing logs (or information), deriving new parameters of interest, and forecasting the level of operational activities. There have been several successful applications of ML in subsurface regression problems. Hampson et al. (2001), Scheutter et al. (2015), Sharma et al. (2017), Bhattacharya et al. (2019), and Zhong et al. (2019) used ML to predict porosity, permeability, TOC, and hydrocarbon production from conventional and unconventional reservoirs utilizing seismic and well log data (Fig. 3.2). Mohaghegh (2017) used ML to predict geomechanical properties in the Bakken Formation, United States.

**Fig. 3.2** An example of a regression problem for permeability prediction using different ML algorithms from a formation in the Appalachian Basin (after Zhong et al. 2019) (Permission granted from SEG)

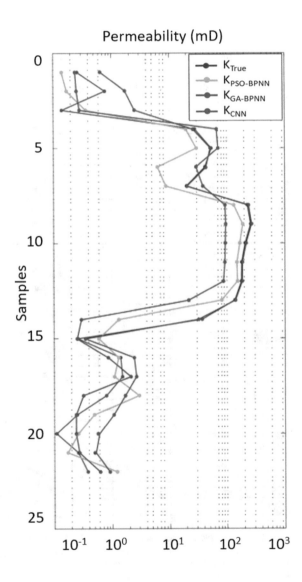

One of the notable differences between classification and regression problems are that the response variables are mostly static in nature at short intervals in classification problems, such as facies. These geologic features do not always change very quickly over recording interval. However, in regression problems, response variables are more dynamic in nature and vary with time and depth; for example, monthly hydrocarbon production. In such cases, we may find that there are both regional (long-term) and local (short-term) trends that control production. This is not very common in classification problems. In some cases, there could be facies associations, which means a set of facies are being repeated due to a particular depositional environment. Sedimentary cycles may coarsen/clean (sand/silt-rich) or fine (clay-rich) upward, form symmetric or asymmetric sequences, or form complete and incomplete cycles, each of which are representative of certain types of depositional conditions. For example, coarsening-upward (or cleaning-upward) sequences are found in progradational deltaic environments (Fig. 3.3). Well logs such as gamma-ray are indicative of shale-sand lithology in this case. However, depending on the frequency of sea-level change and accommodation, there could be several smaller cycles of sediment deposition (or parasequences) that are important to the reservoir development. Changes in low-frequency sequences (e.g., first-order or second-order cycles) is more of a global change over a large geologic time related to eustasy, tectonics, etc., whereas high-frequency parasequences (mostly fourth- and fifth-order cycles) represent changes in

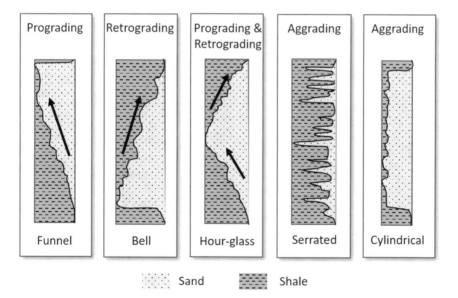

**Fig. 3.3** A simplified example of different types of sequences identified based on well-log motif in a siliciclastic environment (modified after Emery and Myers 1996). The arrows in the sand track indicate coarsening or cleaning upward sequences, whereas the arrows in the shale track indicate fining-upward sequences (Copyright © 1996 Blackwell Science Ltd, Permission received from John Wiley and Sons)

deposition over smaller temporal scales (Fig. 3.4). Please see SEPM (https://www.sepmstrata.org/) and Emery and Myers (1996) for further information on sequence stratigraphy in this context. The bottom line is that the response variables in classification problems can be dynamic in certain cases. There are certain ML algorithms, such as Long Short-Term Memory (LSTM) and Toeplitz Inverse Covariance Clustering (TICC), which are suitable for such dynamic problems. I will discuss these algorithms and their applications in Chapters 4 and 5.

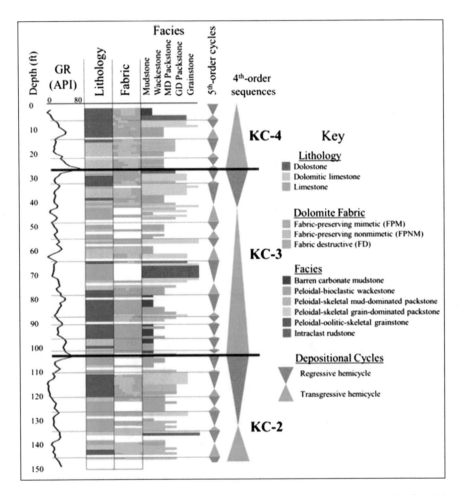

**Fig. 3.4** An example of the cyclic depositional pattern in the Khuff C Formation, Saudi Arabia (Alqattan and Budd 2017). The vertical profile shows the gamma-ray (GR) log, lithology, dolomite fabric, facies, interpreted fifth-order depositional cycles, and fourth-order sequences (AAPG © 2017, reprinted by permission of the AAPG whose permission is required for further use)

## 3.2 Learning Approaches

### 3.2.1 Unsupervised Learning

In an unsupervised learning or clustering approach, we need a dataset that has not been labeled or interpreted yet. That means we do not have the response variable or output assigned to the instances in the dataset. With unsupervised learning, the model classifies the data into relatively distinct groups or clusters based on the natural data distribution and the distance between different groups (Fig. 3.5). Apart from clustering, we also use unsupervised learning for data reduction. In theory, this approach does not require a specific number of clusters; however, there are several software packages in which we need to set up at least the range of possible clusters present in the data. This allows us to visualize the data distribution for different sets of clusters, verify them with our domain knowledge, and select the model that makes most of the sense.

Unsupervised learning is very useful either at the beginning of a project or when we do not have much knowledge about a system yet. It allows for minimum human bias. It does not depend on prior geological knowledge, which is both good and bad, depending on the situations. Unsupervised learning is perfectly applicable to seismic facies classification problems in large areas with 3D seismic data and not enough borehole and outcrop data. In such cases, we can apply unsupervised classification techniques to classify different facies and focus our analysis on the stratigraphic elements (such as point bars, incised valley, shelf-edges, and basin-floor fans) that show the most promising features based on the data and our domain expertise. Coléou et al. 2003 show an example of unsupervised facies classification using the self-organizing map technique (Fig. 3.6). We can also use this technique in cases of well-log-based classification when we do no not have core and cuttings to validate the observations. The same concept applies to remote sensing data.

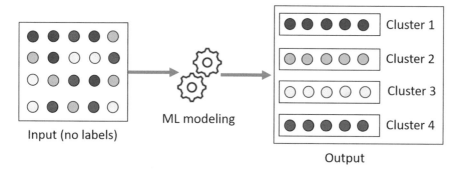

**Fig. 3.5** The concept of unsupervised ML. The original unlabeled data goes into the ML algorithm, which classifies it into different clusters based on user input

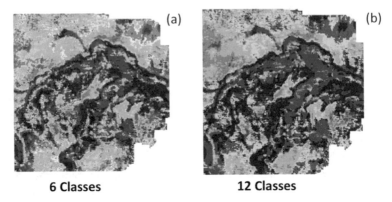

**Fig. 3.6** An example of unsupervised seismic facies classification using the self-organizing map technique (after Coléou et al. 2003). Examples with **a** 6 classes and **b** 12 classes (Permission granted from SEG)

### 3.2.2   Supervised Learning

In supervised learning, we have already assigned the output variables in response to the predictor variables for a portion of the dataset. In this approach, algorithms learn the data pattern or a function that can map the input variables to the output for each instance. Based on the learned function, the algorithms can predict the labels for new and unseen samples with the input features but no pre-assigned labels (Fig. 3.7). This approach is helpful for checking the robustness of the model. In this approach, we

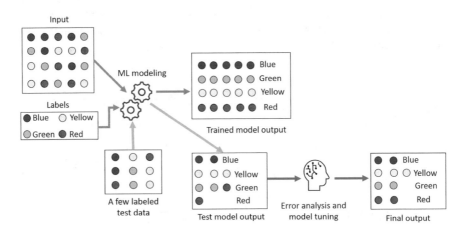

**Fig. 3.7** A simplified concept of supervised ML. Different arrow colors indicate different phases of modeling. The gray arrows indicate the first phase (model training), the green arrows indicate the second phase (model test), and the blue arrows indicate the third phase, in which the model is tuned based on the initial model performance

can use either categorical data (e.g., facies, fractures, faults, etc.) or continuous data (e.g., porosity, permeability, shear wave velocity, etc.) as the output and fine-tune the network hyperparameters based on the results.

Supervised learning is useful when we have some knowledge about the system, but that knowledge does not necessarily cover the whole study area. In such cases, we need a function that can automatically map the input to the desired output in unseen data that does not have labels or interpretations yet. This situation is very common in the appraisal and development stage; often we have enough well logs but not enough core data. We can assign facies based on the core-log integration, apply supervised ML to learn the pattern of the facies associated with different well-log signatures, and use that pattern to predict certain facies in a well not covered by core data (Fig. 3.8). Several researchers have published their work on this problem (Wang and Carr 2013; Bhattacharya and Carr 2019). The beauty of this approach is that it allows us to supervise geologic observations on geophysical, petrophysical, and reservoir responses, building meaningful models that we can use to make actionable decisions. Di et al. (2018, 2019) show an application of supervised deep learning for seismic structural and stratigraphic interpretations and evaluate the model's performance. Supervised learning is also useful in seismic inversion and geologic image classification (such as fossils, minerals, and outcrops).

### 3.2.3  Semi-Supervised Learning

Semi-supervised learning, which lies between completely supervised and unsupervised learning approaches, is a relatively new paradigm in geosciences. Access to ground-truth geologic and reservoir data is limited by cost and logistics, especially in the subsurface. In such cases, semi-supervised or weakly supervised learning approaches can leverage the structure in the unlabeled data to improve classification performance (Karpatne et al. 2017). It allows large volumes of unannotated (or uninterpreted) data to be harnessed in conjunction with typically smaller sets of annotated data (Engelen and Hoos 2020). Figure 3.9 shows the simplified concept of semi-supervised learning. This approach is very useful when working with large datasets in which it is laborious and expensive to label the output for all instances, such as satellite images, 3D seismic data, and micro-CT scans. As geoscientists, we are often asked how we can do more with less. In such cases, a semi-supervised approach could be the answer.

Alfarraj and AlRegib (2019) use a semi-supervised approach for seismic acoustic impedance inversion. In general, deep learning algorithms require a large volume of labeled data for training. Using a semi-supervised learning approach, Alfarraj and AlRegib (2019) could achieve an accuracy of 98% between the estimated and target impedance using 20 seismic traces for training. Dunham et al. (2020) show the applications of two semi-supervised learning approaches in petrophysical log-based carbonate facies classification, which matched or outperformed the fully supervised learning approaches (Fig. 3.10). Figure 3.11 shows the simplified concepts of all

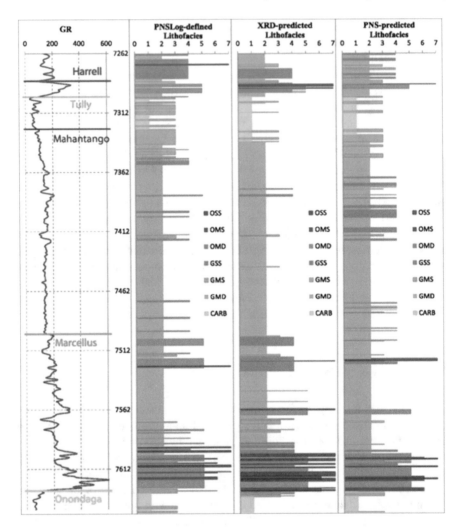

**Fig. 3.8** An example of supervised facies classification in the Devonian interval, including the Marcellus Shale in the Appalachian Basin, United States (after Wang and Carr 2012a). The results are based on an artificial neural network algorithm. The square legends in the log tracks indicate different facies. The results show the overall similarity of facies across the study interval; however, there are sections where the log-predicted facies do not match exactly with the trained facies. See Wang and Carr (2012a) for more details (Reprinted by permission from Springer Nature Customer Service Centre GmbH: Springer Nature, Mathematical Geosciences, Marcellus Shale Lithofacies Prediction by Multiclass Neural Network Classification in the Appalachian Basin, G. Wang and T.R. Carr, © 2012)

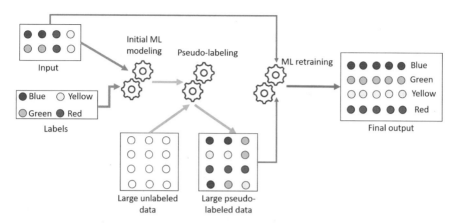

**Fig. 3.9** A simplified concept of semi-supervised learning. Different colors of the arow indicate different phases of modeling. The gray arrows indicate the first phase (model training), the green arrows indicate the second phase (pseudo-labeling), and the blue arrows indicate the third phase (model retraining)

learning approaches, and Fig. 3.12 shows a schematic diagram of the basic workflow used in ML projects. Each of these steps are critical.

## 3.3  Data Pre-Processing

Dataset pre-processing is an important step in constructing a reliable ML model to solve problems. This process involves proper selection of input parameters, which are sensitive to output. Data formatting, cleansing, abstraction, feature engineering, collinearity analysis, and understanding of attribute characteristics are essential steps for the selection of meaningful model input parameters.

### 3.3.1  Data Integration and Feature Selection

After we have formulated the problem and characterized it as either a classification or regression-type, it is time to integrate and select the pertinent data. In our discipline, data come from different sources and in different formats. In addition, we cannot use all the data in our analysis because some of the datasets might not be useful based on our understanding of the physics of the problem. In this stage, we prepare the data for ML modeling.

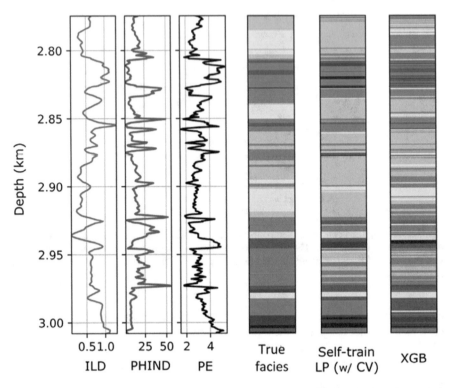

**Fig. 3.10**  A comparison of facies classification by semi-supervised learning (self-train label prop-agation with cross-validation) and XGBoost (XGB) algorithm from a reservoir in Kansas, United States. There are nine facies based on core and log data. Semi-supervised learning with cross vali-dation increased the model performance by ~ 6% compared to the XGB method. See Dunham et al. (2020) for further details (Permission granted from SEG)

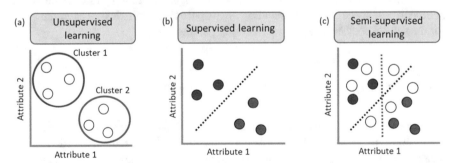

**Fig. 3.11**  The simplified concepts of **a** unsupervised learning, **b** supervised learning, and **c** semi-supervised learning. The unfilled circles indicate unlabeled data, whereas the filled circles indicate labeled data in **a** and **b**. Semi-supervised learning **c** generates pseudo-labels for unfilled circles, which are used in combination with the already labeled samples for ML model building

Processed     ML model      Prediction      Model explanation    Domain expertise    Knowledge presentation
input data                                                       and knowledge
                                                                 discovery

**Fig. 3.12**  A simplified workflow of machine learning modeling and analytics

### 3.3.2  Data Cleansing

Although we may have the pertinent data, it does not mean the dataset is complete for undertaking an ML study. It may contain outliers, null values, missing data, noise, inconsistencies, or formatting issues. Noise can be both random and periodic. For example, well logs can contain varieties of noise in the data due to borehole washout, high-density mud (e.g., barite), improper grounding of electrodes, mechanical failure of the logging tools and assembly, etc. At this stage, we can apply different windowing, filtering, interpolation, and smoothing techniques to remove such effects from data.

### 3.3.3  Statistical Imputation for Missing Data

This is an important step in data pre-processing. Datasets may have missing values, which often causes problems in model performance. While writing a computer code, we can present these values in different ways, such as null, NaN (not a number), N/A, etc. Values could be missing due to several reasons, such as experimental design, bad measurements, access to the system, etc. For example, in petrophysics, we are often not interested in logging the top 50 m (at least) of the subsurface, so to control cost and time, we may not record the conventional logs, except the gamma-ray log. In such cases, we will have missing values in the other logging parameters for the first 50-m interval.

There are a few approaches for imputing missing data. If many values are missing, we could replace them with an indicator variable, keeping in mind whether the missing variables are categorical or continuous in nature. If it is a categorical variable (such as facies), we can simply assign a new category to it. If it is a continuous variable, we can compute mean (known as mean imputation) for those variables missing in a row, if the data distribution is normal. If the data distribution is skewed, we should use median, instead of mean. Replacing a large number of missing continuous variables with a constant value implies that the slope is the same across the interval with missing values, which may sometimes defy geologic rules. For example, consider sonic log data in which the velocity is missing for a few tens of meters along the depth in an overpressured region. We know velocity increases with depth, and in overpressured zones, there would be a drop in velocity. If we do not know exactly where the inflection point of velocity is and we use mean imputation that would be erroneous and could result in wrong decisions in field. We could also carry forward the last response of the recorded variable in the dataset, which is a better method. We could also employ an interpolation strategy of using the last few variables and the variables ahead. We often use this approach in editing well logs. We could also use related data from a similar measurement in an analog field or an offset well in the vicinity. Another less-deployed strategy would be using logic rules. If we can identify when a particular variable is missing in certain situations, we could impute them using logical rules. We can also apply regression-based ML to predict the missing values or well logs from offset wells with good-quality and continuous data. Treatment of missing values is an ongoing area of research.

### 3.3.4   Data Abstraction

In this step, we summarize the data in a format that is usable while keeping the essential features of the data intact. Although ensemble ML models that can combine data in different formats are in the development stage, at this point it is advised to aggregate the data in a similar format for efficient handling and collaboration across platforms. A lot of geodata (such as petrophysical, geomechanical, and temperature logs) are available in ASCII format, whereas the seismic and fiber optic data (distributed acoustic sensing, DAS) are available in SEG-Y format. In general, well log data are sampled at 0.5-ft (0.15-m) intervals, and fiber optic data (DAS) are sampled at a spacing of 2–3 ft (0.6–0.9 m). Bhattacharya et al. (2019) show an example of data abstraction in an unconventional reservoir with a large volume of multi-scale and multi-source data. They show that in such cases, we could divide the well path into small bins, covered by all kinds of pertinent data. We can then compute mean and variance of each parameter in each bin, which can be used for upscaling to a suitable scale of resolution or directly used in modeling.

### 3.3.5   Feature Engineering

Essentially, feature engineering is the process of deriving new features from the original dataset that could be more sensitive to the output. We use the theories of mathematics, statistics, and signal processing to derive new features (Fig. 3.13). While computing new features, we must keep in mind that the new features reveal something extra compared to the original predictor variables. Feature engineering is perhaps the most important step for a successful ML project because physics-based feature engineering can provide us more insight into the data patterns that we can use for diagnostic, predictive, and prescriptive purposes. It is an important step in incorporating domain knowledge to enhance the capabilities of ML models.

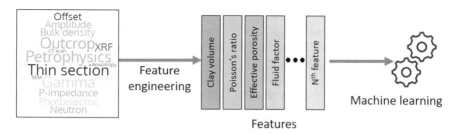

**Fig. 3.13** A visualization of feature engineering in geosciences

Feature engineering helps us in two ways. First, it derives more sensitive parameters that we can use to classify and predict output more successfully. Additionally, the combination of new features to original data can increase dimensionality. Although it is well-known large data dimensionality may reduce the ML model performance, sometimes this may have a positive implication. Wang and Carr (2012a) showed in their study on the Marcellus Shale in the United States that the average distance between different lithofacies clusters can be increased by using feature-engineered parameters instead of a limited number of conventional well logs (Fig. 3.14). Increasing the dimensionality of data can reduce the number of overlapping lithofacies clusters and increase the accuracy of classification.

It is also important to note that certain facies are more sensitive to certain model input parameters, which include both original and derived features. Although

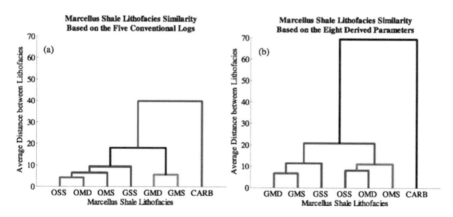

**Fig. 3.14 a** The average distance between different Marcellus Shale lithofacies computed from five conventional wireline logs directly and **b** the eight derived petrophysical parameters. Feature engineering can improve classification results. See Wang and Carr (2012a) for further details (Reprinted by permission from Springer Nature Customer Service Centre GmbH: Springer Nature, Mathematical Geosciences, Marcellus Shale Lithofacies Prediction by Multiclass Neural Network Classification in the Appalachian Basin, G. Wang and T.R. Carr, © 2012)

researchers discuss the curse of dimensionality, there are certain powerful ML algorithms that are fundamentally based on increasing the dimensionality and classification of the data in the feature space. Support vector machine is one such algorithm. At the same time, we should analyze the interdependence among the features, which are common in geosciences. An understanding of feature engineering and its execution can make the ML models more versatile in nature. It is also important to infuse domain expertise and causal understanding while deriving and using feature-engineered attributes in models.

### 3.3.5.1 Data Transformation and Dimensionality Reduction

Data transformation is a common approach of feature engineering. We transform the data into different forms using mathematics and statistics to focus on certain features of the data based on our domain expertise. Geophysicists use feature engineering often. They use different seismic attributes (such as root-mean-square amplitude, frequency, phase, coherence, dip, curvature, flexure, and azimuth, etc.) for enhanced subsurface interpretation. Some of these attributes can be used as predictors for the ML model to classify different structural and stratigraphic features as output (Marfurt 2018; Bhattacharya et al. 2020). Petrophysicists also derive different features from conventional well logs which are very sensitive, such as apparent matrix grain density (RHOmaa), bulk photoelectric absorption (UMaa), and total organic carbon (TOC), to quantify minerals and organic matter. Feature engineering can also include simple data transformations such as the transformation from a linear scale to a logarithmic scale. The resistivity response in rocks and fluid varies significantly ($10^{-2}$ $\Omega$-m to $10^{3}$ $\Omega$-m). Transforming the resistivity log responses from a linear scale to a logarithmic scale can remove the effect of the skewness and improve the model's performance (Wang and Carr 2013; Bhattacharya et al. 2016). We can use these derived petrophysical features in ML projects (Bhattacharya et al. 2016; Bhattacharya and Mishra 2018; Misra et al. 2019). In case of applying ML to sequence stratigraphy, log motifs or patterns are more important than the depth-by-depth values. In such cases, we can derive the convergence of seismic reflectors and the slope of the log curves to use them as input for machine learning. The computation of new features increases data dimensionality. However, some of these features might be redundant, and some of them might not be sensitive to the output. In this step, we also reduce the number of features after analyzing their collinearity. This process enhances the model's generalization capability. We can use principal component analysis, factor analysis, and independent component analysis for dimensionality reduction. Bhattacharya et al. (2019) computed 34 variables from petrophysical, geomechanical, fiber optic, and surface measurements to predict the daily gas production from an unconventional well but used only 18 of them for regression modeling. They found 16 of the derived attributes to be collinear.

### 3.3.5.2   Feature Rescaling

Feature rescaling is an important component of feature engineering. This technique is particularly useful for ML algorithms, many of which are based on distance-based metrics such as K-nearest neighbor (KNN) and support vector machine (SVM), etc. There are several types of feature recalling or normalization approaches, of which two are popular (described below).

Min–Max Scaling

Min–max scaling is the simplest method of data normalization. With this approach, we subtract the minimum value from the variable and divide it over the difference between the maximum and minimum values. By doing this, we standardize the range of the normalized output between zero and one. Petrophysicists often use this method while computing clay volume from gamma-ray logs (Eq. 3.1). The drawback of this technique is that it can include outliers, which can affect the ML results. Therefore, we need to remove outliers from the data before normalization.

$$x_n = \frac{x - \min(x)}{\max(x) - \min(x)} \qquad (3.1)$$

In which $x_n$ is the normalized feature of variable $x$.

Standard Scaling

To reduce the effect of low standard deviation and outliers, we can apply standard scalar (or Z-score normalization) to the data. With this approach, we subtract the mean value from the variables and then divide it over the standard deviation of the distribution (Eq. 3.2). The resulting distribution has a mean of zero and a standard deviation of one. This is a widely used technique in many ML algorithms (Table 3.1).

$$x_z = \frac{x - Mean(x)}{SD(x)} \qquad (3.2)$$

## 3.4   Data Labeling

Data labeling or annotation is another important step, especially in supervised classification. It is highly recommended to work with domain experts at this stage. In supervised learning, we need to label the output in at least the training and validation sets. For classification problems, labeling is binary in nature, which means either

**Table 3.1** An example of feature ($x$) rescaling using min–max scaling ($x_n$) and standard scaling ($x_z$) techniques

| Sample no | $x$ | $x_n$ | $x_z$ |
|---|---|---|---|
| 1 | 10 | 0.00 | $-1.46$ |
| 2 | 15 | 0.13 | $-1.10$ |
| 3 | 20 | 0.25 | $-0.73$ |
| 4 | 25 | 0.38 | $-0.37$ |
| 5 | 30 | 0.50 | 0.00 |
| 6 | 35 | 0.63 | 0.37 |
| 7 | 40 | 0.75 | 0.73 |
| 8 | 45 | 0.88 | 1.10 |
| 9 | 50 | 1.00 | 1.46 |
| **Mean** | 30 | 0.50 | 0 |
| **Standard Deviation** | 13.69 | 0.34 | 1.00 |

the specific label is present or absent. We can label the seismic or petrophysical dataset with different labels of facies, fractures, and faults, which will be used by the ML models to train and test the performance against. For example, if you are working on a fault classification problem, it is suggested working with a structural geologist who has field experiences and an understanding of the mechanical properties of rocks and their interactions with faults. The same applies to working on stratigraphy-related problems.

## 3.5   Machine Learning-Based Modeling

### 3.5.1   Data Splitting

After pre-processing of the dataset, we partition the data into three parts—training, validation, and testing—to implement ML algorithms (Fig. 3.15). In ML, there are several ways to split the dataset. One of the most common approaches is to randomly split the data into training, validation, and testing segments in a specific ratio. The proportion of the dataset in each of these segments is different. In general, we use 60–80% of the data as training and the remaining 20–40% for testing and validation. The question is how we partition the dataset in practice using this method. There are a few strategies. We can compile the whole dataset into one workbook (e.g., Excel™, Access™, Tableau™, etc.) and then randomly divide the data into training, testing and validation segments (i.e., 60-20-20, 80-10-10, 90-5-5).

Here is an example in which we could use a slightly different strategy for the data split. Suppose we are working on a facies classification problem using well logs from 10 wells. We can select six to eight wells for training the model and test the model on the remaining one or two wells. We can show the model performance in

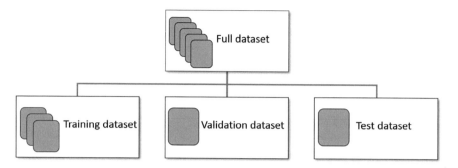

**Fig. 3.15** Data splitting into training, validation, and test segments for ML modeling

both the training and test datasets as a line plot (facies versus depth). In the case of 3D seismic data with 100 seismic sections, we can train the model on 60 sections, validate, and test it on the remaining 40 sections. The advent of deep learning has reduced the number of seismic sections we need for interpretation in the training phase. Bhattacharya and Di (2020) show an example in which they interpreted 30 seismic sections out of 1,000s for fault classification in Alaska using deep learning.

Random partitioning of the data can lead to major issues, such as class imbalance and sample representativeness in the training, validation, and test sets, even when there is no imbalance in the overall dataset (Liu and Cocea 2017). Both issues influence the model performance and its generalization. During the data partitioning process, it is also possible that some of the critical data (e.g., a rare facies or fractures) will fall through the cracks. There are simply not enough samples for these categories. This phenomenon leads to errors in prediction because the model did not have enough training data to understand the relationship between input and output variables. In such cases, we can employ three strategies:

1. Return to the problem and evaluate how important those pre-assigned outputs are. Sometimes, petrophysicists or stratigraphers come up with more than 20 facies in an area integrating core samples, outcrops, and well logs. The question is how many of them are truly important to solving the problem while maintaining scalability and manageability. If some rare facies or features are not important, we can remove them from the dataset.
2. If we think we need to include all labels in the modeling, we can apply statistical techniques to either undersample the majority data or oversample the minority data. This process would balance the dataset so each label will have enough samples to train and test the model.
3. We can apply sophisticated techniques such as synthetic minority over-sampling technique (SMOTE) to randomly generate synthetic data for the minority class to make the dataset more balanced (Chawla et al. 2002). We can either use the nearest neighbors (the number of nearest neighbors to use) or the percentage (the percentage of SMOTE instances to create) to balance the datasets. Bhattacharya and Mishra (2018) used SMOTE to balance their fracture dataset in the

Appalachian Basin, United States, for an efficient performance of ML models. Programing languages and software such as R and Weka have built-in tools for implementing SMOTE.

4.    We can apply intelligent partitioning technique, rather than random splitting, so at least a few samples from each category are present during training and testing processes.

   4a.    One such process is called stratified partitioning. In this process, we keep the distribution of the samples across the classes constant for each parti-tion. For example, if 80% of all samples have label 1 and 20% have label 2, each partition will have the same proportion of samples within its members. This technique gives more reliable results in terms of lower bias and variance.

   4b.    Liu and Cocea (2017) suggested using semi-random partitioning of datasets into training and test segments in granular computing contexts. Granular computing is more like a combination of lumping and split-ting datasets at different levels strategically. This technique involves two operations, such as, granulation and organization. The granulation oper-ation refers to splitting the whole dataset into several subsets (top-down approach), whereas the organization operation corresponds to lumping or merging several subsets into a whole (bottom-up approach). This approach is suitable for balanced and slightly imbalanced datasets; however, it may not be the best method for a highly imbalanced dataset. This is an ongoing area of research.

Cross-validation techniques can be very useful for checking the robustness of the model performance on the whole dataset. In this process, we randomly divide the entire dataset into different subsets or folds (e.g., 5 or 10). Then, we build the ML model, train it on four subsets out of five, and then test it on the remaining subset. After the first round, we train the model on four other subsets and test it on a new subset that was previously used in training. Simply to put, for an n-fold dataset, there will be n iterations; at each iteration, one of the folds is used as the test set, while the remaining $(n - 1)$ folds are used as the training set. This process of training and testing gives more insight into model performance, learnability, and complexity. Figure 3.16 shows an example of five-fold cross-validation.

### 3.5.2  Model Training

After splitting the data into different segments, we build ML models on the training data. We can use any algorithms at this stage, depending on the problem. In the model-building process, we set up the algorithm with optimal network hyperparameters to enable efficient training, validation, and testing. Hyperparameter optimization is a critical step. Each algorithm has different hyperparameters which need to be

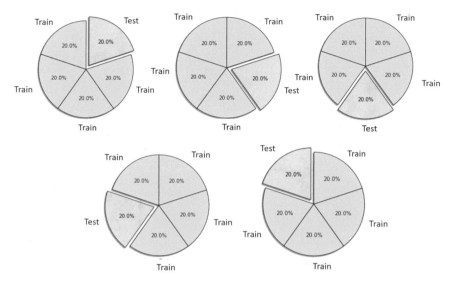

**Fig. 3.16** An example of five-fold cross-validation

optimized. We can optimize the hyperparameters in two ways: grid-search process and trial and error with various combinations of user-defined parameters.

Grid-search process automatically determines the optimized hyperparameters in less time, whereas the trial-and-error approach usually involves more time and cost. Grid search is a systematic approach for fine-tuning network-specific hyperparameters and evaluating the model performance with varying combinations of the hyperparameters specified in a grid. Apart from this method, randomized search or randomized optimization is another method for finding suitable model hyperparameters. The training of the model on the dataset provides a level of accuracy and errors with the original labels or predictions. We can use a confusion matrix or a cross-plot to show the relation between the actual and predicted output. Figure 3.17 shows the concept of grid search and random search in 2D and 3D.

### 3.5.3 Model Validation and Testing

In the next step, we validate and test the trained model. Both validation and test datasets are held back during the model training process. In model validation, we evaluate the performance of the trained model using the validation dataset while keeping the hyperparameters the same. The main objective of this process is to make sure the trained model is generalizable, or not overtrained, in terms of performance. If the model is overtrained, the performance of the model in the validation dataset will be undesirably low. We use different metrics to evaluate model performance. If

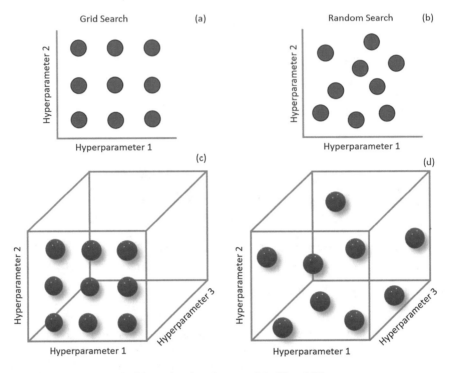

**Fig. 3.17**  The concept of grid search and random search in 2D and 3D

the model is overfit, we will have to go back to the training domain and optimize the network hyperparameters properly.

During validation, we also attempt to understand the complexity of network hyperparameters with model performance, i.e., how the model performance varies (accuracy and error for each label and overall dataset) with network hyperparameters. This process gives insights into the dataset. Some researchers (Mohaghegh 2017) also recommend using a calibration dataset before validation to check the quality and accuracy of results after each iteration.

After model validation, we finally apply the model to the test dataset. The test dataset provides an unbiased evaluation of the final model. Test datasets are only used to assess the model performance, not fine tune the hyperparameters. There is some debate regarding the use of both validation and test datasets in the applied ML community. Many times, the validation set is used as the test set, but this is not recommended. Kuhn and Johnson (2013) propose the use of both validation and test datasets because a test dataset is a single evaluation of the model and it has limited ability to characterize uncertainties in the results from the model. We should also keep in mind that we do not need a validation dataset if we are already using an n-fold cross-validation technique. The validation dataset is applicable when we split the dataset randomly.

# 3.6  Model Evaluation

How can we quantify the performance of models to determine whether they can be used in further analyses? At each stage of model implementation, we must check model performance. That means we must quantify how well the model classifies and predicts output using the training, validation, and test datasets using different metrics (Table 3.2).

## 3.6.1  Quantification of Model Performance and Error Analysis

### 3.6.1.1  Square of Correlation Coefficient

Square of correlation coefficient or the coefficient of determination ($R^2$) is the most popular evaluation metric in both classification and regression models. This measures the strength of the linear relationship between the actual and predicted responses (can be expressed either decimal or percentage). For example, if the $R^2$ value of the classification of a sandstone formation is 0.96 that means the model can explain 96% of variance in the dataset. However, a formation is generally composed of several facies, which stratigraphers can define in various ways.

$R^2$ is a very high-level metric for quantifying model performance. Therefore, we should compute this metric at different granularities, meaning we should look at the accuracies of individual facies in the formation. It is highly likely we will find different $R^2$ values for individual facies in a real-world dataset (including over-prediction and under-prediction for certain facies). There are several other issues with $R^2$. It does not quantify the bias in a dataset (Misra et al. 2019). It does not consider nonlinearity in the data unless we transform the data. It does not provide any understanding of the error at various levels of analysis. Most importantly, it does not indicate causation. This will be discussed more in Chapter 4.

**Table 3.2** Confusion matrices showing actual versus predicted values for a two-class problem. Based on the results, several metrics are computed

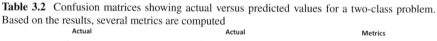
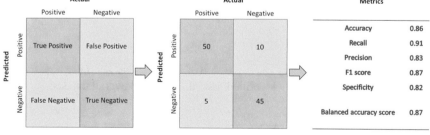

In classification problems, we deal with another dilemma in evaluating accuracy. In any classification problem, we compute the $R^2$ based on the proportion of the predicted values, which are correct out of the total actual observations. Simply put, we compute the total number of instances along the diagonal elements in a confusion matrix and divide that number over the total number of instances to quantify model performance. However, there are several off-diagonal elements, which imply errors in prediction. Based on their nature there are a few other metrics (discussed below) that we should use to assess the overall robustness of the model in prediction.

### 3.6.1.2   Mean Absolute Error

Mean absolute error (MAE) is the average of all absolute errors (Eq. 3.3). Absolute error is the absolute value of the prediction error, which is the difference between the actual value and the predicted value. If there is no error, the difference would be zero; otherwise, absolute error can take either positive or negative values. If we do not take the absolute value, the mean error becomes the mean bias error (MBE), which provides an average measure of the model bias.

$$MAE = \frac{\sum_1^n \left| y_{predict} - y_{true} \right|}{n} \tag{3.3}$$

### 3.6.1.3   Root Mean Square Error

Root mean square error (RMSE) is the square root of the average of the squared differences between actual value and predicted value (Eq. 3.4). RMSE is more like standard deviation. Lower values of MAE and RMSE indicate better model performance with less error; however, there are subtle differences between these two metrics. RMSE gives a relatively higher weight to large errors compared to MAE.

$$RMSE = \sqrt{\frac{\sum_1^n \left( y_{predict} - y_{true} \right)^2}{n}} \tag{3.4}$$

### 3.6.1.4   Recall

Recall is the ratio of true positives to the sum of true positives and false negatives. It is the fraction of the correctly identified instances over the overall dataset.

$$Recall = \frac{True\ Positive}{True\ Positive + False\ Negative} \tag{3.5}$$

### 3.6.1.5  Precision

Precision is the fraction of the true positives over the sum of true positives and false positives.

$$Precision = \frac{True\ Positive}{True\ Positive + False\ Positive} \tag{3.6}$$

### 3.6.1.6  F1 Score

The F1-score is the harmonic average of the precision and recall measurements. The value of the measure varies between zero and one. If the value is zero, this indicates the complete failure of the model, whereas if the value is one, this suggests perfect prediction. In the real-world, we should expect an F1-score somewhere between one and zero.

$$F1\ score = \frac{2 \times Precision \times Recall}{Precision + Recall} \tag{3.7}$$

### 3.6.1.7  Specificity

Specificity is the ratio of the true negatives to the total of true negatives and false negatives.

$$Specificity = \frac{True\ Negative}{True\ Negative + False\ Positive} \tag{3.8}$$

### 3.6.1.8  Balanced Accuracy Score

Balanced accuracy score is the mean of specificity and recall. The value of the balanced accuracy varies between zero and one. In a balanced dataset, this score is identical to accuracy; however, in an imbalanced dataset, this score avoids inflated performance estimates.

$$Balanced\ accuracy\ score = \frac{Recall + Specificity}{2} \tag{3.9}$$

### 3.6.1.9  Kappa Statistic

Cohen's kappa statistic is a measure of the agreement between categorical variables. We compute it from the observed and expected frequencies on the diagonal of a confusion matrix. Some refer to this statistic as a measure of interrater agreement. In general, the value of this statistic ranges from zero to one. A kappa statistic of zero indicates no agreement, 0–0.20 indicates slight agreement, 0.21–0.40 indicates fair agreement, 0.41–0.60 indicates moderate agreement, 0.61–0.80 indicates substantial agreement, and 0.81–1 indicates almost perfect agreement. Kappa statistic may not be useful when working with a skewed or imbalanced dataset.

## 3.6.2  Model Complexity

We consider models to be good based on their predictive performance and their generalization ability with the test dataset. As we quantify model performance, it is also important to realize how model complexities can affect the predictive performance of models. We can analyze model complexity in terms of bias and variance. Figure 3.18 shows the concept of model fitting, variance, and bias. Bias corresponds to the difference between the average of our predicted values from the true mean, and variance represents the scatter (or deviation) of the predicted data from the true mean.

Ideally, we want low bias and low variance in our models, but that is often not attainable with real-world datasets. Hence, there is a trade-off between bias and variance. High bias is indicative of underfitting. Such models are very simple and cannot find the relevant relationship between data points. High variance implies overfitting, which means the model can perform extremely well with the training dataset (low training errors) but underperforms with the independent test dataset (high testing errors). Overfitted models have good memorization ability but not generalization ability, which is more valuable. Figure 3.19 shows model complexity in terms of training and testing errors. See Hastie et al. (2009) for more details on model assessment.

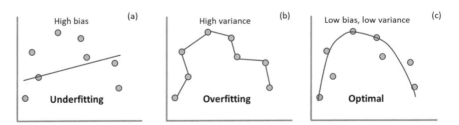

**Fig. 3.18**  The concept of model bias and variance. This is a very useful concept to keep in mind while using machine learning

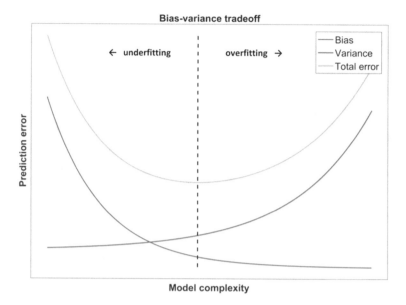

**Fig. 3.19** The bias-variance tradeoff (Kubben et al. 2019), licensed under the terms of the Creative Commons Attribution 4.0 International License (https://creativecommons.org/licenses/by/4.0/). With increased model complexity, bias decreases and variance increases. The dash line indicates reasonable bias and variance, which is suitable for an optimal solution

## 3.7 Model Explainability

Explaining ML model is difficult but important. Often, there is a tug-of-war between model accuracy and model interpretability. Model interpretability is of vital importance because it allows us to improve our trust in the models and identify potential pitfalls in the analysis that we can implement to other areas for new data acquisition and processing. It also helps us in debugging and auditing ML models as needed. This is a growing area of research. We should also keep in mind that model interpretability does not imply causality. Here I describe a few basic methods that we can use for diagnostic and prescriptive analytics. Many of these methods, such as SHAP and LIME, are recently developed tools.

### 3.7.1 Sensitivity Analysis or Key Performance Indicators (KPI)

Sensitivity analysis or key performance indicators (KPI)—also known as variable importance—is critical in data analytics, especially in prescriptive analytics for future work. After we have performed model testing, we need to understand which features

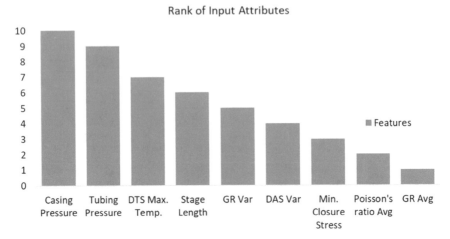

**Fig. 3.20** An example of ranked features used in predicting daily gas production in a hydraulically fractured shale well

are more important than others. This will guide which features we need to care about the most in terms of data acquisition and processing. This step is particularly important at the onset of a large ML project when we may have a plethora of features and want to know which ones to use or remove. There are several techniques for ranking the features, such as fuzzy pattern recognition. Some ML techniques, such as random forest, automatically rank the features in addition to producing model results (Fig. 3.20).

It is important to assess the feature importance provided by ranking algorithms in terms of their physics and reproducibility. We must infuse our domain expertise and causal understanding to interpret the ranks and recommend actions accordingly. Under ideal conditions, when the features are properly and consistently processed and complete datasets exist in all case studies for similar problems (with all labels, in cases of supervised learning), the ranks should be very similar. However, the ranks may differ from data to data for a similar problem, probably due to non-linear feature responses, calibration, scaling, and completeness of the data. Additionally, different ranking algorithms can produce different results. We need to be very careful when assessing the ranks of the features. For example, geomechanical parameters (such as minimum horizontal stress gradient, Young's modulus, Poisson's ratio, etc.) matter significantly to successful production from unconventional reservoirs (Bhattacharya et al. 2019). If we find the surface temperature ranks more than the geomechanical parameters in predicting daily gas production, we need to carefully assess the situation and infer its meaning. Ideally, surface temperature should not have more control on gas production than geomechanical parameters, unless we constrain production due to supply and demand issues for certain months of the year in an area. This is why gasoline prices are generally high in the summer than winter in some countries, including the United States.

Ranks also depend on the geology of a formation. Mohaghegh (2017) shows that the reservoir characteristics of the lower Marcellus Shale have a more distinct impact than those of the upper Marcellus Shale in the Appalachian Basin. Mohaghegh showed that the change in gross thickness and net-to-gross ratio are more important predictors of gas production than initial water saturation and total organic carbon content in the lower Marcellus Shale.

### 3.7.2 Partial Dependence Plots

Partial dependence plots (PDP) show the dependence between the target variable and a set of input features, regardless of other features. There are two types of PDPs: one-way PDP and two-way PDP. Restricting the number of features between one to two helps us understand model complexity. One-way PDPs show the interaction between the target variable and the specific input feature, whereas two-way PDPs show the interactions between two features for a particular target. Figure 3.21 shows examples of partial dependence plots for organic mudstone and calcareous facies in a sedimentary formation.

### 3.7.3 SHapley Additive exPlanations

SHapley Additive exPlanations (SHAP) analysis is a modern tool that can help explain models to an extent (Lundberg and Lee 2017). It is based on game theory. SHAP values help us to analyze feature importance at both global and local scales. SHAP assigns each feature an importance value for prediction. The SHAP value is the average of the marginal contributions across all possible permutations of features, which makes it a more unified approach across global and local scales. However, this also makes SHAP computations very slow. The computational time increases with the number of features used in the model as the algorithm tries to find all possible combinations of features and their contributions. Additionally, the output results from SHAP analysis are approximate solutions. We can use a variety of graphs (e.g., bar plot, beeswarm plot, waterfall plots, decision plots, etc.) to visualize the SHAP values and attempt to explain the model. See https://shap.readthedocs.io/ for further details. Lubo-Robles et al. (2020) showed SHAP's application in classifying salt bodies using 3D seismic data in the Gulf of Mexico, United States. Figures 3.22 and 3.23 show SHAP results from a regression and classification problem.

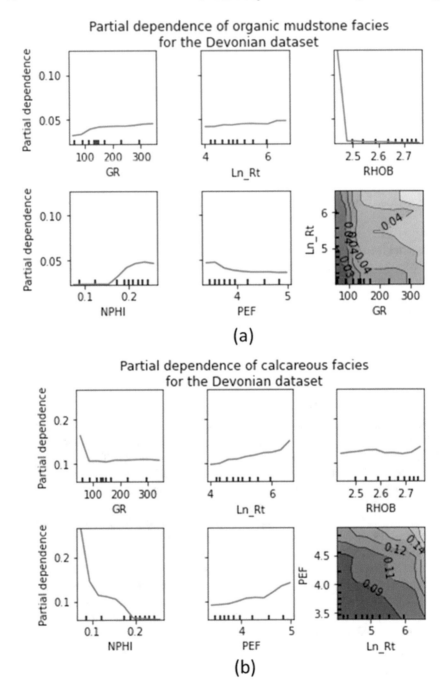

**Fig. 3.21** An example of partial dependence plots to analyze the influence of gamma-ray (GR), logarithm of resistivity (Ln_Rt), and photoelectric factor (PEF) logs on classifying **a** organic mudstone and **b** calcareous facies in a formation. The contour plot in (a) shows organic mudstone corresponds to high GR and high Ln_Rt response, which makes sense because it is radioactive (due to uranium) and resistive (due to kerogen). Similarly, thin carbonate layers exhibit high Ln_Rt and PEF values in **b**

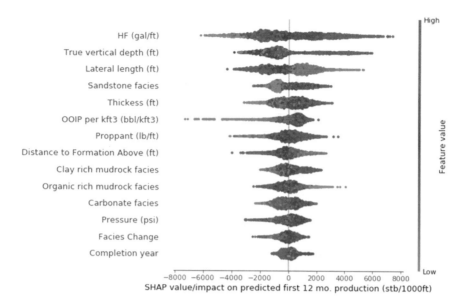

**Fig. 3.22** SHAP analysis for predicting the first 12 months of oil production from a formation in the Delaware Basin, United States (after Yang et al. 2020) (This figure reprinted from Q. Yang, F. Male, S.A. Ikonnikova, K. Smye, G. McDaid, and E.D. Goodman, 2020, with permission from URTeC, whose permission is required for further use)

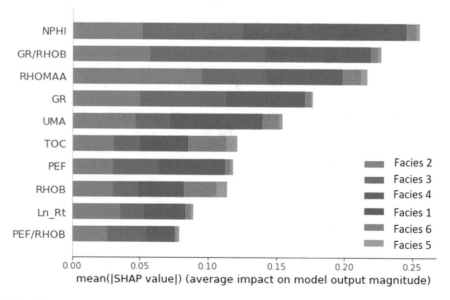

**Fig. 3.23** SHAP analysis for facies classification. Different petrophysical logs have different impacts on facies classification

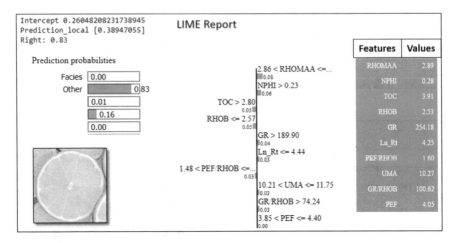

**Fig. 3.24** LIME analysis for an organic-rich mudstone facies classification. Apparent matrix grain density (RHOmaa), neutron porosity (NPHI), and gamma-ray (GR) logs have positive contributions to classifying the facies

### 3.7.4  Local Interpretable Model-Agnostic Explanations

LIME (local interpretable model-agnostic explanations) is a model-independent method that can be applied to any ML models (Ribeiro et al. 2016). LIME helps us understand model behavior and observe how the predictions change by perturbing the input data. This process is similar to the model intervention often used in geophysical inversion processes. ML models at the global level are highly complex and non-linear. LIME assumes we can interpret ML models effectively by using linear relations to approximate local behavior. Unlike SHAP, LIME focuses on training local models to explain individual predictions, not global models. The output from a LIME analysis is a list of explanations showing their individual contributions (Fig. 3.24). In general, LIME is faster than SHAP analysis. In addition to numerical data, LIME also works with image and text data, which is applicable to geology. This method is still in development, and there are some challenges to overcome, such as instability of the model explanations.

Please remember that none of these methods provide causality. We need domain expertise and other techniques, such as Bayesian network theory, regression discontinuity, and factorial design, etc. for that.

## 3.8 Knowledge Discovery, Presentation, and Decision-Making

This is the last step in subsurface geoscience machine learning. At this stage, we visualize and present the knowledge discovered from our data-driven analysis to make decisions on collecting new data, updating models, drilling, and completion.

## References

Al-Anazi AF, Gates ID (2010) Support vector regression for porosity prediction in a heterogeneous reservoir: a comparative study. Comput Geosci 36(12):1494–1503. https://doi.org/10.1016/j.cageo.2010.03.022

Alaudah Y, Michalowicz P, Alfarraj M, AlRegib G (2019) A machine learning benchmark for facies classification. Interpretation 7(3):SE175–SE187. https://doi.org/10.1190/INT-2018-0249.1.

Alfarraj M, AlRegib G (2018) Petrophysical-property estimation from seismic data using recurrent neural networks. SEG Technical Program Expanded Abstracts, 2141–2146. https://doi.org/10.1190/segam2018-2995752.1

Alfarraj M, AlRegib G (2019) Semi-supervised learning for acoustic impedance inversion, SEG Technical Program Expanded Abstracts, 2298–2302

Alqattan MA, Budd DA (2017) Dolomite and dolomitization of the Permian Khuff-C reservoir in Ghawar field, Saudi Arabia. Am Asso Petrol Geol Bull 101(10):1715–1745. https://doi.org/10.1306/01111715015

Bhattacharya S, Carr TR (2019) Integrated data-driven 3D shale lithofacies modeling of the Bakken Formation in the Williston basin, North Dakota, United States. J Petrol Sci Eng 177:1072–1086. https://doi.org/10.1016/j.petrol.2019.02.036

Bhattacharya S, Di H (2020) The classification and interpretation of the polyphase fault network on the North Slope, Alaska using deep learning. SEG Technical Program Expanded Abstracts, 3847–3851. https://doi.org/10.1190/segam2020-w13-01.1

Bhattacharya S, Mishra S (2018) Applications of machine learning for facies and fracture prediction using Bayesian Network Theory and Random Forest: case studies from the Appalachian basin, USA. J Petrol Sci Eng 170:1005–1017. https://doi.org/10.1016/J.PETROL.2018.06.075

Bhattacharya S, Carr T, Pal M (2016) Comparison of supervised and unsupervised approaches for mudstone lithofacies classification: Case studies from the Bakken and Mahantango-Marcellus Shale, USA. J Nat Gas Sci Eng 33:1119–1133. https://doi.org/10.1016/j.jngse.2016.04.055

Bhattacharya S, Ghahfarokhi PK, Carr T, Pantaleone S (2019) Application of predictive data analytics to model daily hydrocarbon production using petrophysical, geomechanical, fiber-optic, completions, and surface data: a case study from the Marcellus Shale, North America. J Petrol Sci Eng 176:702–715. https://doi.org/10.1016/j.petrol.2019.01.013

Bhattacharya S, Tian M, Rotzien J, Verma S (2020) Application of seismic attributes and machine learning for imaging submarine slide blocks on the North Slope, Alaska. SEG Technical Program Expanded Abstracts, 1096–1100. https://doi.org/10.1190/segam2020-3426887.1

Chawla NV, Bowyer KW, Hall LO, Kegelmeyer WP (2002) SMOTE: synthetic minority over-sampling technique. J Artif Intell Res 16:321–357. https://doi.org/10.1613/jair.953

Coléou T, Poupon M, Azbel K (2003) Unsupervised seismic facies classification: a review and comparison of techniques and implementation. Lead Edge 22(10):942–953. https://doi.org/10.1190/1.1623635

Di H, Li Z, Maniar H, Abubakar A (2019) Seismic stratigraphy interpretation via deep convolutional neural networks. SEG Technical Program Expanded Abstracts, 2358–2362. https://doi.org/10.1190/segam2019-3214745.1

Di H, Wang Z, AlRegib G (2018) Seismic fault detection from post-stack amplitude by convolutional neural networks. Conference proceedings, 80th EAGE conference and exhibition, pp 1–5. https://doi.org/10.3997/2214-4609.201800733

Dramsch JS, Lüthje M (2018) Deep-learning seismic facies on state-of-the-art CNN architectures. SEG Technical Program Expanded Abstracts, 2036–2040.

Dunham MW, Malcolm A, Welford JK (2020) Improved well-log classification using semisupervised label propagation and self-training, with comparisons to popular supervised algorithms. Geophysics 85(1):O1–O15. https://doi.org/10.1190/geo2019-0238.1

Emery D, Myers KJ (eds) (1996) Sequence stratigraphy. Blackwell Science, Oxford

Hampson DP, Schuelke JS, Quirein JA (2001) Use of multiattribute transforms to predict log properties from seismic data. Geophysics 66(1):220–236. https://doi.org/10.1190/1.1444899

Hastie T, Tibshirani R, Friedman J (2009) The elements of statistical learning: data mining, inference, and prediction. Springer

Howat E, Mishra S, Schuetter J, Grove B, Haagsma A (2016) Identification of vuggy zones in carbonate reservoirs from wireline logs using machine learning techniques. American association of petroleum geologists eastern section 44th annual meeting. https://doi.org/10.13140/RG.2.2.30165.73443

Huang L, Dong X, Clee TE (2017) A scalable deep learning platform for identifying geologic features from seismic attributes. The Leading Edge 36(3):249–256. https://doi.org/10.1190/tle36030249.1

Karpatne A, Ebert-Uphoff I, Ravela S, Babaie HA, Kumar V (2017) Machine learning for the geosciences: challenges and opportunities. IEEE Trans Knowl Data Eng 31(8):1544–1554. https://doi.org/10.1109/TKDE.2018.2861006

Kubben P, Dumontier M, Dekker A (eds) (2019) Fundamentals of clinical data science. Springer Open. https://doi.org/10.1007/978-3-319-99713-1

Kuhn M, Johnson K (2013) Applied predictive modeling. Springer. https://doi.org/10.1007/978-1-4614-6849-3

Li H, Misra S (2017) Prediction of subsurface NMR T2 distributions in a shale petroleum system using variational autoencoder-based neural networks. IEEE Geosci Remote Sens Lett 14(12):2395–2397. https://doi.org/10.1109/LGRS.2017.2766130

Liu H, Cocea M (2017) Semi-random partitioning of data into training and test sets in granular computing context. Granular Computing 2:357–386. https://doi.org/10.1007/s41066-017-0049-2

Lubo-Robles D, Devegowda D, Jayaram V, Bedle H, Marfurt KJ, Pranter MJ (2020) Machine learning model interpretability using SHAP values: application to a seismic facies classification task. SEG Technical Program Expanded Abstracts, 1460–1464. https://doi.org/10.1190/segam2020-3428275.1

Lundberg S, Lee S (2017) A unified approach to interpreting model predictions. NIPS. https://arxiv.org/pdf/1705.07874.pdf

Misra S, Li H, He J (2019) Machine learning for subsurface characterization. Gulf Publishing

Mohaghegh SD (2017) Shale analytics. Springer

Pires de Lima R, Welch KF, Barrick JE, Marfurt KJ, Burkhalter R, Cassel M, Soreghan GS (2020) Convolutional neural networks as an aid to biostratigraphy and micropaleontology: a test on late Paleozoic microfossils. Palaios 35(9):391–402. https://doi.org/10.2110/palo.2019.102

Ribeiro MT, Sameer S, Guestrin C (2016) "Why should I trust you?": Explaining the predictions of any classifier. Proceedings of the 22nd ACM SIGKDD international conference on knowledge discovery and data mining, pp 1135–1144. https://doi.org/10.1145/2939672.2939778

SEPM Strata (2020) Cycles in the stratigraphic record. https://www.sepmstrata.org/Terminology.aspx?id=cycle

Scheutter J, Mishra S, Zhong M, LaFollette R (2015) Data analytics for production optimization in unconventional reservoirs. SEG Global Meeting Abstracts, 249–269. https://doi.org/10.15530/urtec-2015-2167005

Sharma R, Chopra S, Lines L (2017) A novel workflow for predicting total organic carbon in a Utica play. SEG Technical Program Expanded Abstracts, 1887–1891. https://doi.org/10.1190/segam2 017-17735087.1

Van Engelen JE, Hoos HH (2020) A survey on semi-supervised learning. Mach Learn 109:373–440. https://doi.org/10.1007/s10994-019-05855-6

Wang G, Carr TR (2012a) Marcellus Shale lithofacies prediction by multiclass neural network classification in the Appalachian basin. Math Geosci 44:975–1004. https://doi.org/10.1007/s11 004-012-9421-6

Wang G, Carr TR (2012b) Methodology of organic-rich shale lithofacies identification and prediction: a case study from Marcellus Shale in the Appalachian basin. Comput Geosci 49:151–163. https://doi.org/10.1016/j.cageo.2012.07.011

Wang G, Carr TR (2013) Organic-rich Marcellus Shale lithofacies modeling and distribution pattern analysis in the Appalachian Basin. Am Asso Petrol Geol Bull 97(12):2173–2205. https://doi.org/10.1306/05141312135

Wu X, Liang L, Shi Y, Fomel S (2019) FaultSeg3D: Using synthetic datasets to train an end-to-end convolutional neural network for 3D seismic fault segmentation. Geophysics 84(3):IM35–IM45. https://doi.org/10.1190/geo2018-0646.1

Yang Q, Male F, Ikonnikova SA, Smye K, McDaid G, Goodman ED (2020) Permian Delaware Basin Wolfcamp a formation productivity analysis and technically recoverable resource assessment. SEG Global Meeting Abstracts, 561–570. https://doi.org/10.15530/urtec-2020-3167

Zhao T (2018) Seismic facies classification using different deep convolutional neural networks. SEG Technical Program Expanded Abstracts, 2046–2050. https://doi.org/10.1190/segam2018-2997085.1

Zhong Z, Carr TR, Wu X, Wang G (2019) Application of a convolutional neural network in permeability prediction: A case study in the Jacksonburg-Stringtown oil field, West Virginia, USA. Geophysics 84(6):B363–B373. https://doi.org/10.1190/geo2018-0588.1

# Chapter 4
# A Brief Review of Popular Machine Learning Algorithms in Geosciences

**Abstract**  In the last several decades, computer scientists and statisticians have developed and implemented a plethora of machine learning (ML) algorithms. Although the application of data-driven modeling is relatively new to geoscience, we can trace back some of its early applications to the 1980's and 1990's. This chapter will discuss the fundamental theory and analytic framework of many popular ML algorithms. Understanding the fundamentals of these algorithms, network-specific hyperparameters, and their meaning is essential to better implement these algorithms in our datasets and enhance the success rate of data-driven modeling. These algorithms are based on solid mathematical and statistical theories. Indeed, some algorithms are better than others for certain types of applications; however, sometimes, our lack of understanding of algorithms and the nuances of their applications to specific datasets cause them to underperform compared to others. Once we understand the fundamentals of algorithms and our datasets, ML will be more fun and provoking, which will facilitate further progress of geo-data science.

**Keywords**  Machine learning algorithms · Model hyperparameters · Clustering · Neural networks · Decision trees · Deep learning · Ensemble learning · Physics-informed machine learning

## 4.1  K-means Clustering

K-means is undoubtedly one of the most popular algorithms in unsupervised classification or clustering. It is conceptually simple, easy to implement, and versatile in nature. Most of the statistics software and scripting languages have some version of in-built tools for K-means clustering. The origin of K-means can be traced back to 1950–1960s (Steinhaus 1956; MacQueen 1967). The original paper by Steinhaus is written in French, and the conclusions closely resemble the modern K-means architecture. However, it is widely believed that James MacQueen used the term 'K-means' for the first time.

K-means is a partition-based clustering method. It attempts to partition N-dimensional datasets into k sets (or clusters) based on a sample (MacQueen 1967).

Each cluster has a centroid; therefore, this algorithm assumes the original data distribution inside each cluster as spherical or circular around the centroid. The word "means" in K-means refers to averages. We start the modeling with a pre-defined number of clusters. It clusters the data in such a way, so the data points inside one cluster are similar to each other, whereas they are dissimilar to other data points in other clusters (Fig. 4.1). Therefore, the variance tends to be low within a cluster and high outside it. The clustering is primarily based on a distance-based metric that is used to determine the similarity between data points and assign them to different clusters. We can use different types of distance measures, such as Euclidean and Mahalanobis. It assigns data points to a cluster such that the sum of the squared distance (SSE) between the data points and the cluster's centroid (or arithmetic mean of all the data points in that cluster) is at the minimum (equation below). Then, the K-means method recomputes the centroids by taking the mean of all data points belonging to the same cluster. This step is critical as the centroids computed in the first step are random and may not be the accurate centroids of the data belonging to each cluster. The repositioning of cluster centroids with further iterations reduces the variance inside the clusters, as they become close to the 'true' centroids of each cluster. We can run this process several times until the solutions are converging or matching with our perceived targets. We need to control a few parameters in K-means, such as the number of clusters, method for initialization, distance metric, and the number of iterations.

$$SSE = \sum_{i=1}^{k} \sum_{x_j \epsilon P_i} \|x_j - c_i\|_2^2 \tag{4.1}$$

Where x indicates data points inside k number of P clusters, $\|.\|_2$ indicates Euclidean L2 norm, and c corresponds to cluster centroids.

How do we select the number of clusters in K-means? We can use scatterplots, pairplots, and a priori geologic knowledge (if any) to understand the number of possible clusters to start with. For better quantification, we can use the elbow method and silhouette analysis. In the elbow method, we apply K-means using different

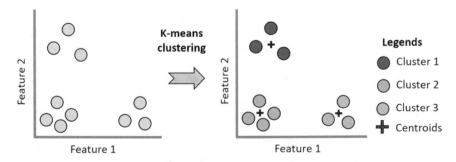

**Fig. 4.1** A simplified concept of K-means clustering

cluster numbers and then plot the number of clusters versus the sum of squared distance between data points and their assigned clusters' centroids. Figure 4.2 shows a decreasing error with the increase in the number of clusters. The optimal number of clusters is selected from that point or area where the error starts flattening out. Sometimes, it is hard to identify the optimal cluster number as the plot may be monotonically decreasing and not show a distinctive elbow. Another technique is using silhouette analysis. This method computes silhouette coefficients of each data point. It quantifies how much a data point is similar to its own cluster compared to other clusters. The value of the coefficient ranges between −1 and 1. A value of the coefficient close to '1' indicates the sample in one cluster is far from other clusters, whereas a value close to '−1' indicates the samples might have been assigned to the wrong cluster. If the coefficient equals zero, the sample is very close to other neighboring clusters. We would want the coefficient value as close to one as possible. Figure 4.3 shows the results from the silhouette method to identify the optimal number of clusters.

It is also important to realize our data contains multiple types of information, including facies, fractures, faults, and rock properties. Some of the attribute expressions of these geologic features may seem very similar. The random initialization in K-means may lead to a risk of mixing the target features with similar ones into the same cluster (Di et al. 2018). Therefore, we should use our geologic knowledge to understand the geologic meaning of the clusters. In terms of initialization, there are different methods, such as Forgy (1965), K-means++ (Arthur and Vassilvitskii 2007), and principal component analysis-part (Su and Dy 2007), etc. While using K-means, it is highly recommended to use standardized data; otherwise, the results

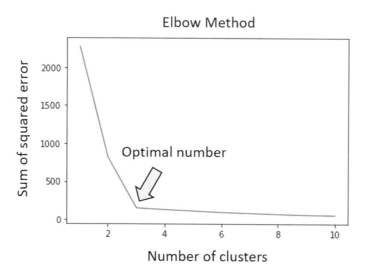

**Fig. 4.2** An example of the elbow method showing the optimal number of clusters is three

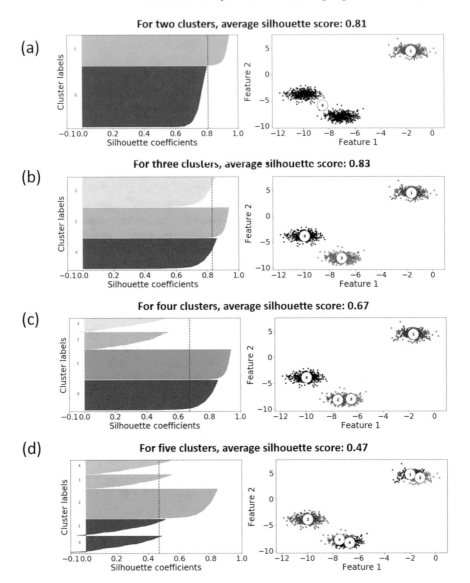

**Fig. 4.3** An example of silhouette coefficients with respect to different cluster numbers. The optimal cluster number is three

can be misleading due to the domination of variables with larger variances compared to other variables. It is highly applicable to geophysical and petrophysical attributes.

Most often, K-means method does not provide global optimal solutions. Due to its gradient descent nature, it can converge to a local minima rather than global minima (discussed further in the artificial neural network below). It can also be affected by noise, outliers, and varying data density that can drag the centroids to wrong positions, thereby increasing the variance. See Celebi et al. (2012) for further details on the advantages and disadvantages of K-means clustering. Regardless of some of its drawbacks, K-means clustering has been widely used in geosciences (Coléou et al. 2003; Matos et al. 2007; Al-Mudhafar and Bondarenko 2015; Di et al. 2018).

## 4.2   Artificial Neural Network

Artificial neural network (ANN) is perhaps the most well-known ML algorithm used in scientific disciplines. In essence, ANNs attempt to mimic how the human brain processes information and yields results (McCulloch and Pitts 1943; Bishop 1995; Kordon 2010).

ANN is primarily composed of three layers: input, hidden, and output (Fig. 4.4). These three layers are connected via neurons, which transport the information from one layer to the next. We feed the original input features to the input layer, which distributes them to the hidden layer. The hidden layer is the key component of the ANN structure. It learns data structure in terms of patterns and interrelationships among the input features, and then it distributes the learned data patterns (mathematically expressed as weight) to the output layer (Bhattacharya et al. 2016). An activation function controls the output of a node. Activation functions work as a switch, meaning certain outputs are generated when certain relationships between the input parameters are found. There are several activation functions, such as sigmoid, hyperbolic tangent (tanH), rectified linear unit (ReLu), etc. We must use nonlinear activation functions to introduce nonlinearity to ANN, otherwise, the output becomes a simple linear function, which is not encountered in many real-world problems.

Suppose we are working on a simple supervised rock type (e.g., sandstone and shale) classification problem using gamma-ray (GR) and neutron porosity (NPHI)

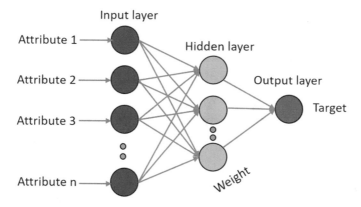

**Fig. 4.4** The architecture of an artificial neural network composed of an input layer, hidden layer, and output layer. This is an example of a feed-forward neural network

logs using ANN. The hidden layer learns the relationship between the GR and NPHI logs and how sandstone and shale are assigned in the training dataset. Based on physics, shale exhibits higher GR and higher NPHI values than sandstone due to radioactivity and clay-bound water. In this case, the shale's output nodes will be activated if the instances in the dataset show high GR and high NPHI responses; otherwise, the nodes for sandstone will be activated as the output.

In general, there are two types of ANN: feed-forward and back-propagation. Feed-forward ANN is a simple type of ANN. It can be either a single-layer perceptron or a multi-layer perceptron (MLP). In a feed-forward ANN, neurons' connections among different layers do not form a cycle or loop; the information flows forward (input layer → hidden layer → output layer). ANN modeling starts with a randomly assigned weight, then a set of patterns is repeatedly fed forward, and then the weights of the neurons are modified until the output values match the actual values (Bhattacharya et al. 2016).

Why should we care about weights? It is because some inputs are more important than others to yield output. Weights represent the strength of these inputs (Mohaghegh 2017). In the case of a backpropagation neural network (BPNN), the output is compared to pre-assigned output (the training dataset), and the error (or the difference) is then propagated backward to adjust the weight of the neurons (Fig. 4.5). This process continues iteratively until we reach a satisfactory level of convergence between the pre-assigned output and the BPNN-derived output. BPNN utilizes mean-squared error and gradient descent methods to update the weight of the neurons.

What is the impact of epoch? Epochs refer to the number of cycles the full training dataset is passed through the network, and iteration is the number of batches or steps to complete one epoch. Geophysicists and petrophysicists are very familiar with iterative processes from dealing with inversion modeling. Increasing the number of

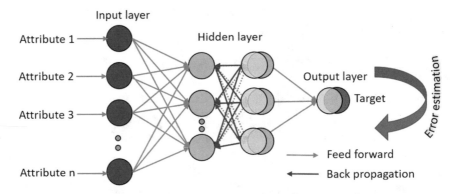

**Fig. 4.5** The architecture of a back propagation neural network. In such networks, the error between a pre-assigned target response (red circle) and a model-based target response (yellow circle) is propagated backward (blue arrows) to adjust the weight of the neurons (yellow circles)

epochs can lead to neural network memorization, which may make it overtrained. An overtrained neural network is as useful as no model.

How do we know that a neural network model is overtrained? If the training model performance is 100% and the test performance far less than that, the model is not generalized. For example, the $R^2$ for the training model can be close to 100%, but somewhere near 30–50% for the test set. In such cases, we need to make the model more generalized. We should keep in mind that the issue of model overfitting is not just limited to neural networks. Several other ML algorithms are highly affected by this phenomenon.

The obvious question is how we avoid overtraining a network. We need to update the model hyperparameters and carefully assess the input features. In the case of ANN, there are several other hyperparameters to control while designing the ANN model, such as the number of hidden layers, number of hidden layer nodes, learning rate, damping coefficient or momentum, activation functions, and number of epochs for better optimization. Let us learn about these hyperparameters in more details.

## 4.2.1 Hidden Layer

We have already defined hidden layer. Based on experience, simple classification and regression models can be handled with just one or two hidden layers with different nodes. For image classification problems, I recommend using a deep neural network with several hidden layers.

## 4.2.2 Learning Rate

The learning rate hyperparameter corresponds to how quickly a neural network model can learn the data patterns. The value of the learning rate ranges from zero to one. If the learning rate is very small, the network operates slowly, and the neurons' weight coefficients are updated slowly. If we set learning rate to a higher value, the network will learn quickly, but too high of a value will make it more unstable, and it may get stuck at local minima. If the model gets stuck at the local minima, the model performance cannot improve further (Fig. 4.6). Therefore, this hyperparameter must be optimized. With a suitable learning rate, the loss function (or error) decays with the number of epochs. The nature of decay determines the optimal value of the learning rate. The optimal learning rate depends on the topology of the loss function. Figure 4.7 shows the scenarios with different learning rates.

Although statistical software packages and programming languages provide default values for this hyperparameter, we can fine-tune it. Smith (2018) suggests starting the modeling with a low learning rate and increasing the rate at each iteration. Smith proposed using cyclical learning rates (CLR) and learning rate range test (LR range test) to find the optimal learning rate. If we plot the learning rate and

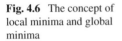

**Fig. 4.6** The concept of local minima and global minima

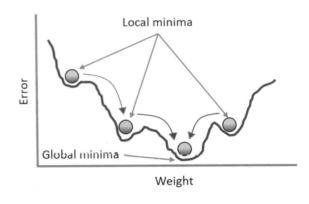

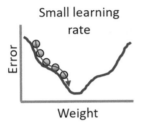

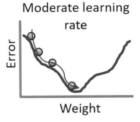

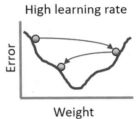

**Fig. 4.7** Scenarios with different learning rates. A small learning rate delays the model convergence, whereas a high learning rate make the model unstable and unpredictive

loss function with the number of iterations, we will find that at one point (or a small region), the loss function shows the steepest descent, which after a while may get unstable for higher learning rates. The value where we observe the steepest descent is the optimal value of the learning rate.

## 4.2.3 *Momentum*

Momentum accelerates the training process. The utilization of momentum in ANN modeling helps to stabilize the test dataset. We must optimize this hyperparameter along with the learning rate. Similar to the learning rate, the momentum value ranges from zero to one. Small values of momentum can dampen the training process. If we assign a small momentum value, the model cannot avoid local minima, and it may get stuck there. Increasing the value of momentum increases the size of the loss function's steps by trying to jump from local minima toward the global minima and makes the training process faster, which is desirable. The higher the value of momentum, the higher the possibility that the network can escape the local minima and reach the global minima. Figure 4.8 elucidate the concept of momentum.

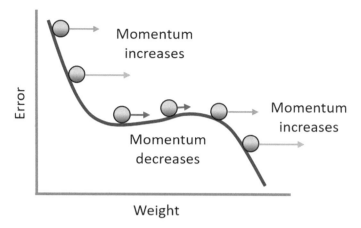

**Fig. 4.8** A simplified concept of momentum. The arrows with different lengths indicate increasing or decreasing momentum, depending on their instantaneous position. Green arrows indicate high momentum on the steep gradient, whereas red arrows indicate slow momentum on the flat gradient

For optimal ANN modeling, momentum and learning rate should have an inverse relationship. If we assign a large value to momentum, we may want to keep the learning rate smaller. If we assign large values to both momentum and learning rate, skipping the global minima with huge steps is possible. In such a case, the loss function will be divergent, and test results will be unstable. It is suggested keeping a lower learning rate and higher momentum in commonly encountered problems in subsurface geosciences.

## 4.2.4 Activation Function

Fine-tuning the activation function can change ANN model performance. Figure 4.9 shows a variety of activation functions. Over the last several decades, either sigmoid or tanH functions have been used with considerable success. However, these activation functions are suitable for traditional ANNs with one hidden layer, not for deeper networks, as these functions saturate quickly at their minimum and maximum values, introducing a vanishing gradient problem (discussed further in "Recurrent neural network and long short-term memory," below). In ANNs with many hidden layers, the gradient diminishes drastically as it is propagated backward through the network. By the time the error in the loss function reaches the layers, it becomes so small that it may have minimal effect. In such cases, the network becomes unstable. We can use either ReLu or leaky ReLu functions in these scenarios. ReLu functions are becoming quite popular in multilayer perceptrons and convolutional neural networks, which will be discussed later. It is also possible to assign different activation functions for different hidden layer sets, enhancing model performance.

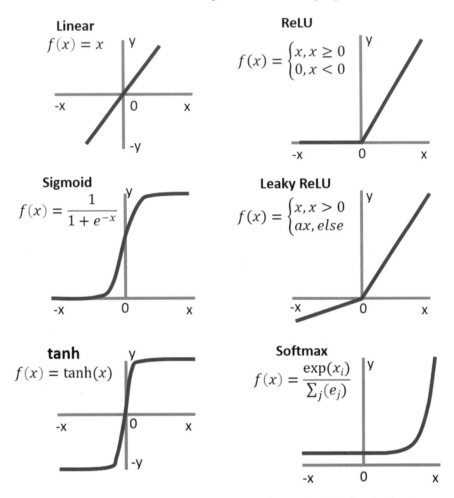

**Fig. 4.9** Varieties of activation functions used in neural networks. Each function has its own advantages and disadvantages

## 4.3   Support Vector Machine

Support vector machine (SVM) is a popular ML algorithm. Unlike ANN, SVM has only recently found applications in subsurface geosciences (Kuzma 2003; Al-Anazi and Gates 2010; Wang et al. 2014; Bhattacharya et al. 2016; Misra et al. 2019). Vapnik developed the SVM algorithm in the 1990's using the kernel trick, which is based on a solid mathematical background of statistical learning theory (Cortes and Vapnik 1995; Christianini and Shawe-Taylor 2000; Kordon 2010). We use SVM in both classification and regression problems. In the case of regression problems, we refer to them as support vector regression (SVR). SVM requires fine-tuning of at least two hyperparameters, including penalty and gamma, in the case of radial basis functions, which I will explain below.

In theory, SVM maps the original data from the input space to a higher dimensional (or even infinite-dimensional) feature space so that the distance between each data point increases and classification of different variables into classes becomes simpler (Luts et al. 2010). Figure 4.10 shows the concept of SVM. We use a kernel function for high-dimensional mapping. Essentially, these functions are projection or mapping functions. Because we cannot perceive the data in a higher-dimensional feature space, these functions help us see the results.

Support vectors are key features of SVM (Fig. 4.11). In the case of a simple binary classification problem, these vectors are the data points (i.e., samples) which lie on the boundaries of different classes (e.g., facies and fractures) during classification (Bhattacharya et al. 2016). There can be many hyperplanes, which can distinguish two classes. The SVM algorithm finds the optimal hyperplane, which is the farthest from both classes. For binary classification problems, SVM assumes two planes that

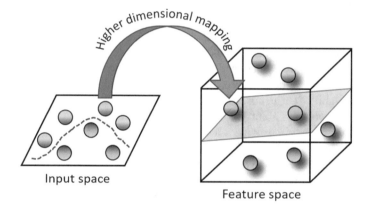

**Fig. 4.10** The concept of higher-dimensional mapping in support vector machines (modified after Kordon 2010) (Reprinted/adapted by permission from Springer Nature Customer Service Centre GmbH: Springer Nature, Machine Learning: The Ghost in the Learning Machine by AK Kordon © 2010)

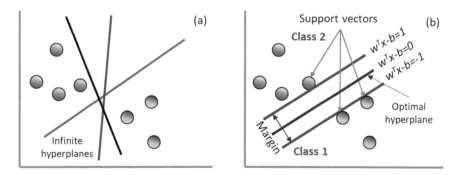

**Fig. 4.11** A support vector machine for classification

support each class and maximizes the distance (also called "margin") between them. The optimization problem involves pushing these parallel planes or support vectors apart until they collide with each class's data points. As we perform the classification, there is still a chance that data points might be overlapping. In such cases, the data points are not easily separable. We use the soft margin concept to penalize the data points on the wrong side of the margin. This penalty parameter (also called "C" or "empirical error" in SVM) allows a limited number of misclassifications to be tolerated near the margin (Mishra and Datta-Gupta 2018). A larger C value assigns a higher penalty to errors.

There are several kernel functions available, such as linear, polynomial, radial basis function (RBF, also known as Gaussian kernel), sigmoid, and mixture. These functions must meet Mercer's conditions. Table 4.1 shows their mathematical expressions. It is important to understand some of the different capabilities of kernel functions in terms of their interpolation and extrapolation abilities. We commonly use RBF and polynomial kernels in complex problems. RBF has one parameter that we need to control: gamma. Smaller gamma values reduce the model's ability for interpolation, whereas higher gamma values increase its interpolation ability. This also depends on how close the data points are.

In the case of a polynomial kernel, the ability to interpolate and extrapolate data depends on the polynomial degree. In general, a higher degree will have improved interpolation ability at the cost of extrapolation, whereas a lower degree will have increased extrapolation ability. We should keep in mind that no single parameter of a kernel function will provide a model with both interpolation and extrapolation properties (Zhong and Carr 2016). In such circumstances, we can use a mixture kernel (i.e., a mixture of polynomial and RBF kernels) to preserve both properties and provide a more robust model. How can we find a suitable kernel function and penalty parameter? It depends on the data. Ideally, we should perform a grid-search method to find the optimal values (Fig. 4.12).

**Table 4.1** Table showing different types of kernels

| Kernels | Expressions | Parameters |
|---|---|---|
| Linear | $K(x_i, x_j) = x_i.x_j$ | |
| Polynomial | $K(x_i, x_j) = (x_i.x_j + 1)^d$ | d: degree of polynomial |
| Radial Basis Function | $K(x_i, x_j) = (\exp(-\Upsilon\|x_i-x_j\|^2)$ | $\Upsilon>0$ |
| Sigmoid | $K(x_i, x_j) = \tanh(m(x_i,x_j)+b)$ | m (slope) and b(intercept) |

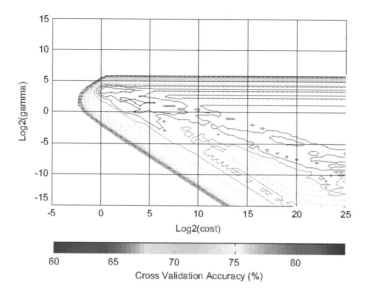

**Fig. 4.12** An example of grid search used in support vector machine (Wang et al. 2014) (Reprinted from Computers & Geosciences, 64, G Wang, TR Carr, Y Ju, and C Li, Identifying organic-rich Marcellus Shale lithofacies by support vector machine classifier in the Appalachian basin, 52–60, Copyright (2014), with permission from Elsevier)

Can we apply SVMs to multi-class problems? SVMs were primarily developed for two-class problems; however, there are several successful case studies of implementation of SVM with multi-class problems. SVM handles multi-class problems efficiently with one-against-all and pairwise (one-against-one) comparison techniques (Hastie and Tibshirani 1998). The pairwise comparison method leads to more straightforward binary classification than the one-against-all method (Wang et al. 2014).

There are a few advantages of using SVM. SVMs derive solutions from a significantly small portion of the dataset (Pal and Foody 2012); this capability is critical in subsurface evaluation, where seismic, well log, and core data are costly and limited. An advantage of SVM over ANN is the need to optimize only a few user-defined hyperparameters, such as penalty parameter and kernel function. Moreover, SVM-derived solutions are global solutions. Therefore, there is no chance of the local minima problem ANN encounters.

As far as disadvantages, SVM model development is computationally intensive. This affects the model performance and increases the cost of computation in large datasets with many features (Pal and Foody 2010; Bhattacharya et al. 2019). Because large, complex, and nonlinear datasets require many support vectors to separate them, they run up a computational cost. Essentially, the ratio of the training samples to the data dimensionality decreases for many support vectors, making parameter estimation inaccurate at that level. This may result in poor performance of SVM compared to other algorithms (Bhattacharya et al. 2019). SVM is sensitive to the choice of good kernel functions and penalty parameters. A single kernel function may not always be helpful. Zhong and Carr (2016) show a successful application of mixed kernels that can overcome the limitations of individual kernels in specific scenarios (Fig. 4.13). There are also limited infrastructure and commercial statistical software packages available for SVM processing due to its high complexity and

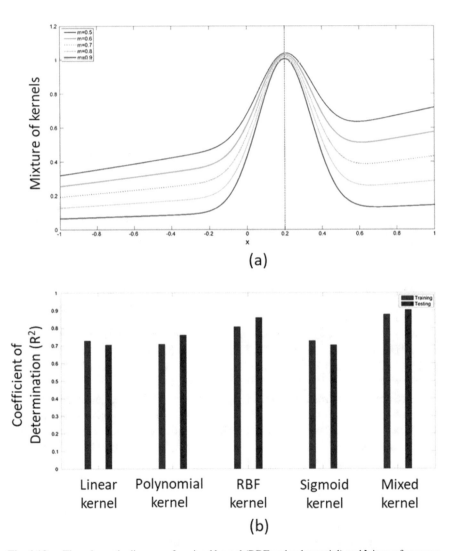

**Fig. 4.13** **a** The schematic diagram of a mixed kernel (RBF and polynomial) and **b** its performance over other kernels in predicting reservoir oil minimum miscibility pressure (Zhong and Carr 2016) (Reprinted from Fuel, 184, Z Zhong and TR Carr, Application of mixed kernels function (MKF) based support vector regression model (SVR) for $CO_2$ – Reservoir oil minimum miscibility pressure prediction, 590–603, Copyright (2016), with permission from Elsevier)

recent arrival. However, we can overcome this by using open-source languages, such as Python and R.

## 4.4  Decision Tree and Random Forest

### 4.4.1  Decision Tree

The decision tree is a very well-known and easy-to-understand technique used in both classification and regression problems. Belson described the first decision tree (DT) in 1959. This algorithm produces a tree-like structure, which resembles a flowchart. The trees are composed of three parts: the decision node, branches, and leaf node (Fig. 4.14). DT starts at the root node (the topmost node) and ends at several leaf nodes. It learns to partition the data based on certain conditions of the input features in a recursive manner. Branches represent the chance outcomes connecting the root nodes and internal nodes. DTs select a feature at each node and classify the data into two groups based on certain thresholds.

We can think of the operation of DTs in terms of if–then conditional statements, commonly used in programming. For example, if condition 1 and condition 2 prove true, then outcome A occurs; otherwise, outcome B occurs. An example in petrophysics would be if gamma-ray response is high and total organic carbon content is high; it is probably an organic-rich mudstone, otherwise, it is an organic-lean rock.

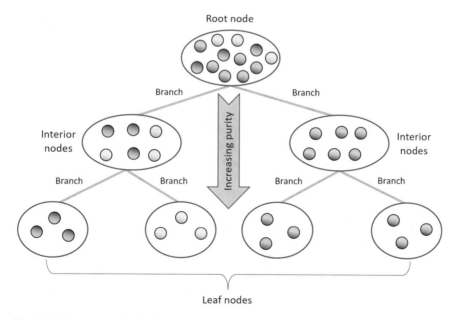

**Fig. 4.14** The concept of a decision tree

If we want to further classify whether the organic-lean rock is sandstone, limestone, or dolostone, we can add a few more criteria, such as density, neutron porosity, and photoelectric factor logs to further characterize these rocks. The DT method finds the threshold of the features using the criteria for the maximum drop in impurity— measured via metrics—such as information gain, Gini index, and gain ratio. The branches form a hierarchical structure.

The features that are most sensitive in terms of their impact on dropping the impurity or entropy are at the upper level, followed by features of relatively less importance. The dataset continues to split into smaller subsets if either the resulting nodes are purer than the previous node or some stopping criteria is met. If the tree becomes very large, the resultant leaves are pure, close to 100%. Such trees may have issues in managing and generalizing the training model and the tree becomes overtrained, making it unable to successfully implement the test dataset.

To create useful DT-based models, we need to set up a few network hyperparameters. The most critical hyperparameters include the stopping criteria, maximum depth, minimum number of samples for each split, minimum sample leaf size, maximum features at each split, etc. The maximum depth refers to the depth of the tree. In practice, we may limit the maximum depth of the tree to prevent overfitting. If we assign infinite or no values in the maximum depth, the tree will expand until all leaves are pure or until all leaves contain less than the minimum number of samples per split. The good thing about this is the model becomes more complex and can capture more information about the dataset, which could include heterogeneities and nonlinearities. However, as the tree grows large and attempts to produce the leaves as pure as possible, the model becomes overtrained and cannot be generalized for the test datasets. If we assign a very low value to maximum depth, then the model does not accurately learn the data patterns and becomes underfit. In general, increasing the maximum depth decreases bias and increases the variance of the model. Modules in programming languages (such as Scikit-learn in Python) find the maximum depth value for a given dataset automatically; we can then optimize these values based on training and testing scores and errors.

We must keep in mind that changing one hyperparameter affects the others. The minimum number of samples to split an internal node and the minimum number of samples required to be at a leaf node are required to ensure that the samples' adequate number drives every decision in the tree at each node (external or internal). We can vary these hyperparameters between one sample and all samples at each node. If there are not adequate samples to drive the decisions for splitting the samples, the model becomes overfit because it becomes so dependent on the specific features of the available samples that it fails to generalize. If we assign a high number of samples, then the model may become underfit. In general, increasing these hyperparameters increases bias and decreases the variance of the model. See chapter 3 for more details on model fitting. The maximum number of features refers to the number of features to consider when looking for the best split. Changing the number of maximum features can change the model performance to an extent (Bhattacharya and Mishra 2018), as shown in Fig. 4.15. In general, decreasing the maximum features increases bias and reduces variance.

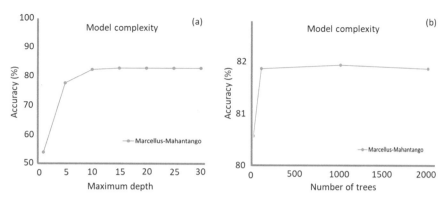

**Fig. 4.15** An example of model complexity (after Bhattacharya and Mishra 2018) showing **a** the impact of the maximum depth on model performance and **b** the influence of the number of trees on model performance. It is interesting to see the model performance flatten after reaching a certain level of complexity (Reprinted from Journal of Petroleum Science and Engineering, 170, S Bhattacharya and S Mishra, Applications of machine learning for facies and fracture prediction using Bayesian Network Theory and Random Forest: Case studies from the Appalachian basin, USA, 1005–1017, Copyright (2018), with permission from Elsevier)

We can use a few metrics to evaluate the quality of DT models. Information gain or entropy is the measure of the drop in the input dataset's impurity. It is the difference between impurity before split and average impurity after split. The gain ratio is the ratio of the information gain over the split information. This ratio helps reduce bias for attributes with several outcomes. The attribute which shows the highest ratio is selected as the splitting attribute.

Gini is a simple statistical measure of the distribution in a population used to infer the inequalities. In ML, this index measures the probability of a random variable being incorrectly classified. The value of this index ranges from zero to one. If the value is zero, it indicates all variables belong to a particular class, and if it is one, then the variables are distributed across several classes. A Gini index of 0.5 represents the equal distribution of variables across classes. We select the features with the minimum Gini index as the splitting attribute.

### 4.4.2 Random Forest

The random forest (RF) algorithm is an ensemble of classification trees that can be used in classification and regression problems using a majority voting scheme (Breiman 2001) (Fig. 4.16). RF starts with a set of classification trees, each created from random subsets of input data consisting of input and output variables (Mishra and Datta-Gupta 2018). Unlike DTs, the RF algorithm starts the model training process with many decision trees in parallel with bagging (or bootstrap aggregation). Each decision tree in RF contains information about the random subsets of the full

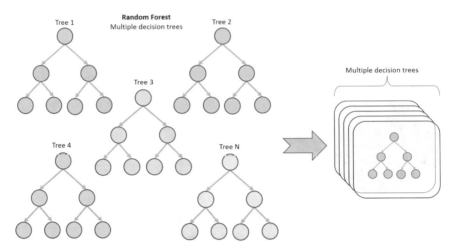

**Fig. 4.16** A simplified concept of random forest, which is a combination of multiple decision trees

dataset. At the end, RF aggregates all this information from the decision trees by averaging, which reduces variance in the model. For prediction purposes, RF uses the "out-of-bag" samples because each tree uses only a subset of the dataset. For the mathematical principle of random forest, please see Cutler et al. 2011.

RF needs several network hyperparameters, such as the maximum depth, predictor variables at each node, maximum features, and the total number of trees. Bhattacharya and Mishra (2018) and Bhattacharya et al. (2019) show the impact of hyperparameters on RF model performance. Using a large subsurface dataset from the Appalachian Basin, they show that the model's accuracy increases as the depth of the trees (maximum depth) and the number of trees increase; however, the accuracy flattens out (or saturates) after reaching a certain level. It is also essential to keep in mind that increasing these two hyperparameters will increase the computational cost, which is not always favorable. As per the number of trees in the RF, Holdaway and Irving (2017) suggested using hundreds of decision trees for predictive modeling in subsurface applications. Increasing the number of trees in RF can improve model stability (Liu et al. 2017). Liu et al. (2017) also suggest that an extreme tree depth can decrease the model's stability, whereas a shallow depth can undesirably underfit the model.

Apart from its application in classification and regression models, RF can be used to analyze the importance of predictors built into it. This is a unique feature of this algorithm. For classification problems, measures such as mean decrease impurity (MDI) or Gini importance and mean decrease accuracy (MDA) or permutation importance are used to rank the input features (Breiman 2001; Louppe 2014). RF considers the increase or decrease of impurity with the changes in the input features to rank them.

Unlike DTs, RF has low variance and bias. RF is an ensemble of several decision trees containing information of different subsets of the full data, the results of which are aggregated into the final result. This reduces the DT's overfitting problem and error due to variance (Louppe 2014). However, RF can suffer from an overfitting problem if the underlying decision trees have a very high variance, which could be due to significantly high depth and a low minimum number of samples per split. Many decision trees in a random forest can also reduce the error due to bias to an extent (Bhattacharya and Mishra 2018).

RF is better than DT in terms of improved model performance. However, the interpretability of the RF model could be an issue. DT-based models are very fast and inexpensive to build. They are very useful for visualizing and explaining relationships in the data. We can use DT over RF when we want a simple model to explain the nature of relations, do not have enough computational power, and are not concerned about high accuracy. However, DTs suffer from high variance factor, which is not an issue with RF.

## 4.5  Bayesian Network Theory

Bayesian network (BN) is another widely known algorithm in statistical analysis. BN has been applied to several subsurface case studies. The concept of BN is based on conditional probability. For a well-log-based facies or fracture classification problem, we can mathematically express Bayesian network theory as:

$$P(f|l) = \frac{P(l|f)P(f)}{P(l)} \tag{4.2}$$

which $P(f|l)$ represents the posterior probability of the target or class ($f =$ facies or fracture) given the predictor ($l =$ well logs), $P(f)$ represents the prior probability of a class, $P(l|f)$ indicates the probability of the predictor given class information, and $P(l)$ is the prior probability of the predictor.

Directed acyclic graphs (DAG) are perhaps the most useful graphical representations of the BN theory. DAGs can show the direction of causality in a Bayesian network. We can define the structure of a DAG in terms of nodes (random variables) and arrows (Ben-Gal 2007). The nodes represent input features and output, whereas the arrows connecting them represent the direct connection or dependence between two variables. The input features are considered the parents, and the output is considered the descendant. Each node in the input has a conditional probability table that quantifies the effects the parents have on the node. Figure 4.17 shows an example of DAG in a Bayesian network problem in which three input parameters can model an output. The presence of an arrow between two nodes indicate one influences the other because one of them is the parent in this case, whereas the absence of an arrow indicates no direct relationship between them (Hernán and Robins 2006; Thornley et al. 2013). Not all input features may be connected via arrows; they may not be directly related to each other, but they can influence the output. We should also note that the arrows among the input parameters and output do not form a directed cycle, so the

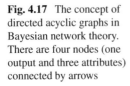

**Fig. 4.17** The concept of directed acyclic graphs in Bayesian network theory. There are four nodes (one output and three attributes) connected by arrows

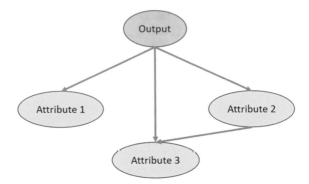

graph is a proper DAG. If properly used, DAGs can reveal a lot more about the data patterns that may not be easily comprehensible by traditional statistical measures. Of course, we do need to interpret DAGs with our domain expertise. For a detailed mathematical treatment of causality, see Pearl (2009).

In theory, all possibly important variables are identified in a fully causal DAG, and each variable is completely defined in terms of all possible states (Huang et al. 2008). Such a BN structure is ideal for causal analysis by domain experts. This may be possible as a project matures with more data and interpretations.

The complexity of BN depends on the number of parents and the local score metric. If the number of parents in BN is one, DAGs will not show any connections among the input variables. This implies that the input features are independent (Fig. 4.18). Although this condition simplifies the mathematical problem, many features in the subsurface are interrelated. Therefore, setting the number of parents to one does not reveal the true complexity of the dataset. Increasing the number of parents may reveal complex relations between input features and output. Bhattacharya and Mishra (2018) also found that increasing the number of parents after a certain level might not increase the model performance (Fig. 4.19), implying saturation of the model. A few input features may also remain disconnected because they are truly unrelated.

The other important aspect of BN theory is related to the network's mode of learning the network structure. There are various approaches to structural

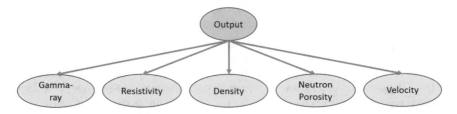

**Fig. 4.18** An example of a Bayesian network with one parent, which implies all attributes are independent. This is not always the case, especially in petrophysics. Increasing the number of parents will provide more insights

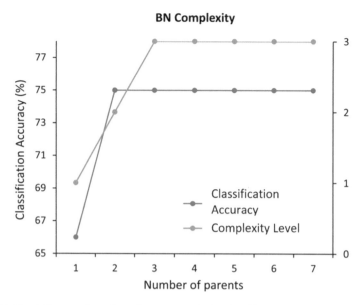

**Fig. 4.19** The influence of number of parents on facies classification accuracy (modified after Bhattacharya and Mishra 2018) (Reprinted from Journal of Petroleum Science and Engineering, 170, S Bhattacharya and S Mishra, Applications of machine learning for facies and fracture prediction using Bayesian Network Theory and Random Forest: Case studies from the Appalachian basin, USA, 1005–1017, Copyright (2018), with permission from Elsevier)

learning, such as constraint-based algorithms, score-based algorithms, and hybrid. The constraint-based approach is inspired by the seminal paper published by Pearl and Verma (1991). It uses conditional independence tests to learn the dependence and independence structure of the data. Score-based approaches use goodness-of-fit scores as objective functions to maximize (Cooper and Herskovits 1992; Heckerman et al. 1995; Bouckaert 1995, 2008; Scutari et al. 2019). This is another optimization problem in BN; otherwise, the model will tend to be overfit (Carvalho 2009). We can use scoring functions such as minimum description length (MDL), Akaike information criterion (AIC), Bayes Dirichlet likelihood-equivalence uniform joint distribution (BDeu), Bayes, etc. as local score metrics. The penalization function built into these metrics attempts to simplify the DAG structure and avoid overfitting. Bhattacharya and Mishra (2018) show the comparison of different local scoring functions on model performance in facies classification (Fig. 4.20). They show that the Bayes- and entropy-based scoring functions outperform other approaches such as MDL, AIC, and BDeu. A hybrid approach combines both constraint-based and score-based approaches.

Naïve Bayes is a commonly used technique in Bayesian network theory which assumes the conditional independence of the input features. Although this could be useful in certain cases, it is not at all helpful to understanding the complexity of the dataset.

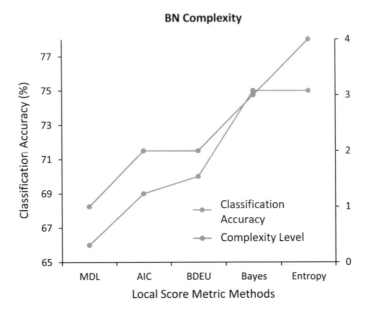

**Fig. 4.20** The influence of number of parents on facies classification accuracy (modified after Bhattacharya and Mishra 2018) (Reprinted from Journal of Petroleum Science and Engineering, 170, S Bhattacharya and S Mishra, Applications of machine learning for facies and fracture prediction using Bayesian Network Theory and Random Forest: Case studies from the Appalachian basin, USA, 1005–1017, Copyright (2018), with permission from Elsevier)

So far, I have discussed traditional ML algorithms. Although these algorithms are useful, they are not as powerful for analyzing computer visions or dealing with large image and sequence datasets. In such cases, we implement deep learning algorithms. In essence, deep learning algorithms have deep internal structure, which facilitates the analysis of large datasets. In the next section, I will discuss a few deep leaning algorithms.

## 4.6   Convolutional Neural Network

Convolutional neural network (CNN) is the most widely known deep learning algorithm. Recently, it has emerged as the go-to algorithm for image classification in geoscience, especially for facies and fault classification problems. By using a CNN on a benchmark dataset, He et al. (2016) showed an error of only 3.57% compared to human-driven classification with an error rate of 5.1% (Russakovsky et al. 2015) on the same benchmark dataset. Several researchers used deep learning algorithms for automated structural and stratigraphic feature classification (Di et al. 2018; Dramsch and Lüthje 2018; Zhao 2018; Wu et al. 2019; Alaudah et al. 2019). In general, there are two types of CNN architecture: fully connected CNN (FCN) and encoder-decoder type. Both architectures have been used in geosciences, depending on the problem and granularity.

## 4.6.1  Fully Connected Network

Similar to traditional ANN, FCN is composed of several layers, including a convolutional layer, pooling layer (or subsampling layer), and fully connected layer (Fig. 4.21).

### 4.6.1.1  Convolutional Layer

The convolutional layer is the first unit in an FCN. We feed the input data to the first convolutional layer, and it extracts features from the input images. Computers read images as pixels, which we can express as the matrix x × y × z (height by width by depth/channel). For a black-and-white image, z = 1, and for an RGB image, z = 3. Next, we convolve a kernel function (or a filter) with the original image to produce feature maps (Figs. 4.22, and 4.23). Feature maps are the output of the convolution process. These maps can help us understand how CNN generates the features used in modeling. What are the implications of the feature maps? The feature maps that result from CNN are not same as geologic maps. They cannot be readily interpreted by human eyes, but they do represent patterns to computers. This is an ongoing area of research. We perform the convolutional operation by sliding the kernel over the input data. This process is followed in multiple steps.

In the first step, the kernel function is convolved with a portion of the input data having the same dimensions as the kernel function. This window is often called the receptive field. After the first convolution, the pixel value of the original image in the receptive field is multiplied by the pixel value in the kernel, which is stored in the feature map at the exact location. It is at this stage (just before generating the feature

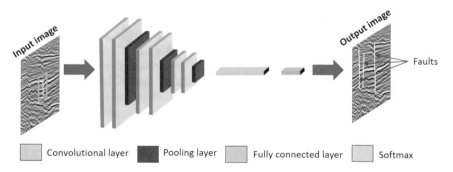

**Fig. 4.21**  An example of a fully connected convolutional neural network

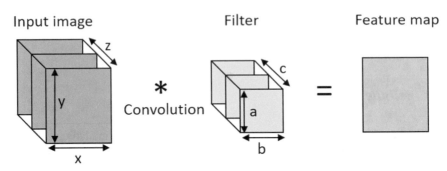

**Fig. 4.22** The concept of convolution to generate feature maps in a convolutional neural network

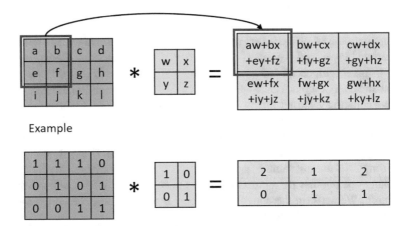

**Fig. 4.23** The concept of convolution in a convolutional neural network with an example. The 3 × 4 matrix on the left is the input image, which is convolved with a 2 × 2 filter to produce the 2 × 3 feature map. The stride here is one

map) when we apply an activation function (i.e., ReLu function) to introduce non-linearity to the CNN model. The output of the basic convolution operation passes through the activation functions and is stored in the feature map.

In the next step, the kernel moves to the next receptive field, convolves with the original image at that position, and generates another set of values stored in the feature map and corresponding to that location. This process continues until all the pixels in the original image are covered by the kernel function. We aggregate and store all the values generated from this convolutional process in the final feature map. The complexity of the feature map increases with the number of convolutional layers. The initial feature map generated after the passing the image through the first convolutional layer is simpler and closer to physically interpretable features.

Why don't we use a kernel function whose dimension is the same as the original image? We could, but the output results will be blurred. Using a kernel function with a significantly smaller size than the full original image helps preserve the granularity

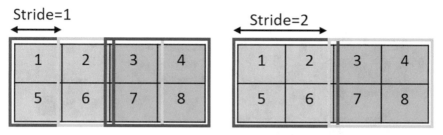

**Fig. 4.24** An example of different strides (one and two) on a 2 × 4 input image

of the information. A feature map's size depends on three factors: depth, stride, and padding (Loussaief and Abdelkrim 2018). Depth corresponds to the number of filters used for the convolution operation. Stride is the number of pixels to skip by the filter (or kernel) over the input matrix (Fig. 4.24). It reduces the input image's size, which we can express using the formula.

$$\frac{n + f - 1}{s} \tag{4.3}$$

In which $n$ is input dimensions, $f$ is filter size, and $s$ is stride length. By default, the value of stride is one. Zero-padding refers to the addition of zeroes in columns and rows in the input matrix. The padding ensures information at the borders is retained.

### 4.6.1.2 Pooling Layer

After generating the high-dimensional feature map, we use a pooling layer to reduce the number of features and complexity of the model. Essentially, the pooling layer merges several semantically similar features into one. This process reduces training time and variance and enhances the training model's generalization capability by reducing the chance of overfitting.

One of the most common methods is maximum pooling. Maximum pooling extracts the most important features (essentially, the maximum values), such as edges. It retains approximately one-fourth of the whole dataset (Fig. 4.25). Although we use a pooling layer in most cases, the nature of the problem and data availability should indicate to us whether a pooling layer is needed or not. For example, if we are working with a small set of borehole geophysical logs or geochemical data, the feature map produced by the convolutional layer is not significantly high-dimensional with several features. In such cases, using a pooling layer will reduce the model performance, which is undesirable. This is of particular importance when we have a statistically rare output (i.e., a particular facies) that is critical to the geologic analysis. However, the use of a pooling layer is generally recommended when working with large datasets, such as 3D seismic and fiber-optic data.

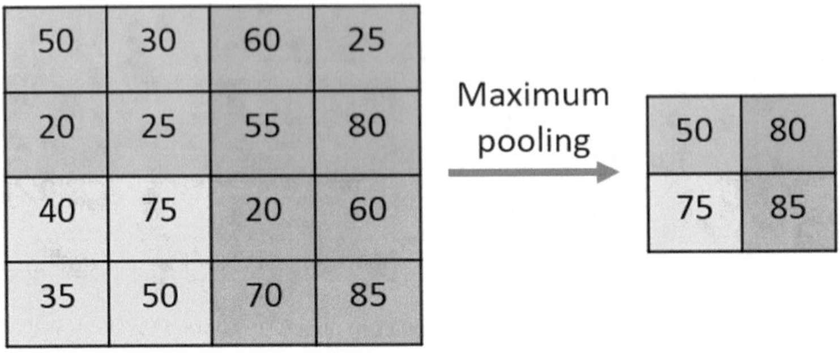

| 50 | 30 | 60 | 25 |
|----|----|----|----|
| 20 | 25 | 55 | 80 |
| 40 | 75 | 20 | 60 |
| 35 | 50 | 70 | 85 |

Maximum pooling →

| 50 | 80 |
|----|----|
| 75 | 85 |

**Fig. 4.25** An example of 2 × 2 maximum pooling

### 4.6.1.3   Fully Connected Layer

In the next step, the final pooled feature map is fed to the fully connected layers. Fully connected layers generate the final output (Fig. 4.26). These layers receive the previous layers' output and flatten them to transform them into a single vector. The output can be discrete or continuous in nature, depending on the problem (i.e., classification and regression).

## 4.6.2   Encoder-Decoder Network

Because of the issues related to pooling layers in FCN, which results in blurred output with localization problems, some researchers prefer using an encoder-decoder network (Badrinarayanan et al. 2015). Encoders detect and classify objects, whereas decoders locate objects in the image accurately. The encoder's role is to encode the

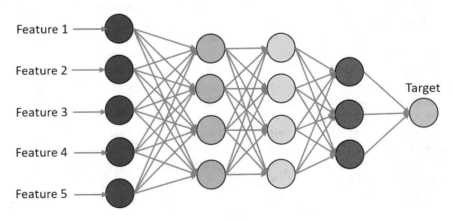

**Fig. 4.26** An example of fully connected layers in CNN

input data into a feature vector that captures its semantic information, which is then passed into the decoder that generates the best possible match to the actual or intended output. These networks are popular in natural language processing. Recently, several researchers have used this technique to classify faults, facies, channels, salt bodies, etc. using seismic data (Alaudah et al. 2019; Di et al. 2018; Pham et al. 2019; Sen et al. 2019; Zhang et al. 2019).

Encoders contain stacks of convolutional layers with batch-normalization and activation functions, followed by pooling layers (Fig. 4.27). Please note that the fully connected layer is removed from the encoder. This makes the network significantly smaller and easier to train (Badrinarayanan et al. 2015). Generally, the decoder component contains stacks of deconvolution layers and unpooling layers. The deconvolution layers attempt to recover feature maps at the original size, recovering the spatial dimensions. This is also known as the semantic segmentation. Both deconvolution (or transpose convolution) and unpooling layers facilitate upsampling.

We can design an encoder-decoder network in different ways using hyperparameter optimization strategies. In general, a deep encoder-decoder network can outperform a shallow one to an extent; however, it depends on the problem and computational cost. In a conventional U-net or encoder-decoder architecture, each decoder block receives input from the corresponding encoder block. However, Sen et al. (2019) show a modification in which the decoders can also receive input from the encoder blocks below it. This puts constraints on the upsampling operation. We can also optimize the stride in the decoder layer. A smaller stride implies reconstruction of more details in the output.

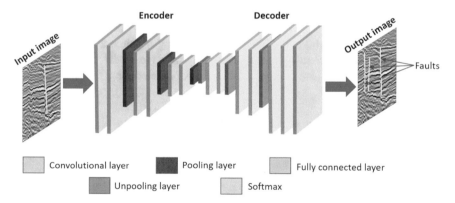

**Fig. 4.27**  An example of an encoder-decoder neural network

### 4.6.3   Optimizing CNNs

There are several hyperparameters to fine-tune in CNN. We can classify these hyper-parameters into two types: spatial feature learning and training. Table 4.2 shows some of these hyperparameters (Mboga et al. 2017). Some researchers keep the training hyperparameters constant while varying the spatial feature learning hyper-parameters; however, there are several studies in which both types are optimized. For optimization, we use grid search, random search, weighted random search, genetic algorithm, etc. (Hinz et al. 2018; Andoine and Florea 2020).

Recent studies in geosciences show the popular use of two or more convolutional and pooling layers. So, how many layers do we need to use? It depends on the complexity of the problem and the dataset fed to the model. Multiple convolutional layers increase the number of features extracted from the data to analyze the complex problem. In contrast, multiple pooling layers reduce model complexity and prevent it from being overfitted. We can keep the size of each convolutional layer the same or we can vary them. Wu et al. (2019) shows an example of six convolutional layers, in which every two consecutive layers (i.e., first and second, third and fourth, fifth and sixth layers) have the same dimensions. In general, deep neural networks with several convolutional and pooling layers provide good results. However, we may lose the object's location-related information as the network becomes deeper (Alaudah et al. 2019). We must be careful to check the degree of information loss.

### 4.6.4   Strategies to consider in CNN Modeling

Because ML-based analysis, especially deep learning requires a large amount of data, we can apply different strategies to properly train the model, described below. Some of these strategies are highly applicable to CNN modeling.

#### 4.6.4.1   A Large Volume of Annotated Data

Fully annotated datasets are critical to building a good CNN model. This is an ideal approach in which domain experts use their knowledge to interpret a portion of the subsurface data to be used to train the model. For example, a structural geologist would know more about faults, fractures, and their relationship to rock rheology and tectonics than a computer scientist would. Similarly, a stratigrapher would be more knowledgeable about the stratigraphic sequences and underlying first-order to fourth-order causal mechanisms, such as global sea-level change, changes in local flow conditions, accommodation space, etc. In such cases, we can use either patch-based analysis or section-based analysis in CNN.

Patch-based analysis is based on training the model on randomly selected patches extracted from the input data (i.e., inline and crosslines in a 3D seismic survey).

**Table 4.2** CNN hyperparameters

| CNN hyperparameters | Definition and function | Common range of values in literature |
|---|---|---|
| *Spatial feature learning hyperparameters* | | |
| Number of convolutional layers | Convolutional layer extracts features from the input original data and feature maps | 2–8 |
| Number of fully connected layers | Generates the final output | 1–3 (2 is most common) |
| Number of filters | An increase in the number of filters increases the number of features learned but may cause overfitting | 16–1024 |
| Filter size | A matrix of weights used to convolve with the input data input. Smaller filters are used in local feature detection, whereas larger filters are used in global feature detection | $2 \times 2$–$11 \times 11$ |
| *Training hyperparameters* | | |
| Learning rate | How quickly a neural network model can learn data patterns? We can use SGD, Adam, and Adagrad optimizers. In addition to learning rate, we also use a decay function | 0–1 |
| Momentum | Accelerates the training process and helps stabilize the test dataset | 0–1 |
| Maximum number of epochs | The number of times the training dataset passes through the network | 10–1,000 |
| Early stopping criteria | The termination of training before the model uses many epochs and becomes overtrained. This is done by computing a score function after each epoch and assessing whether the network outperforms the previous best model | Depends on the data, less than the maximum number of epochs |
| Weight decay | A regularization technique used to prevent the weights from growing too large and causing overfitting | 0–0.1 |

(continued)

**Table 4.2** (continued)

| CNN hyperparameters | Definition and function | Common range of values in literature |
|---|---|---|
| Dropout rate | Temporary disabling of neurons aimed at reducing overfitting. This allows neurons to operate independently by reducing the network's overdependence on a small number of neurons | |

In section-based analysis, we annotate an entire section or multiple sections in the image and test the remaining sections. It also allows us to capture the relation between stratigraphy, structure, and time/depth along a few sections and classify those learned patterns in the test dataset. Section-based approaches would be useful to fault classification if their distribution covers the full dataset. With this approach, it is generally hard to miss the overall data pattern, unless there are local changes, such as rotational faults and variable offset. In such cases, it is advised to use 3D cubes of annotated datasets, rather than 2D sections for CNN training because the expression of many geologic features such as faults changes along different azimuths. Therefore, we need to visualize the data first to understand the geologic features in the study area and then decide whether a patch-based approach is more beneficial than a section-based approach, depending on the resources and deadlines.

### 4.6.4.2   Annotated Public Data

The amount of publicly available subsurface data has been sluggishly increasing in the recent years that we can use to test models, for example, SEG SEAM dataset and the F3 3D seismic dataset in the Netherlands. These datasets contain certain structural and stratigraphic features (e.g., folds, faults, clinoforms, etc.) that are present in several other areas in the world. We can use such annotated public data for training and testing models with our own dataset. However, we must be careful with annotations, feature scaling, signal-to-noise ratio, etc. for public data. This approach is not recommended if we are not convinced of the relation between the geology and the geophysical and petrophysical features between the two datasets.

### 4.6.4.3   Synthetic Data

Using synthetic data is a very efficient and popular approach in the geophysics community because manual annotation of a real dataset is expensive in terms of time and cost. We can generate synthetic seismic and well-log data using fundamental principles of geophysics and signal processing. With this approach, we can generate thousands of synthetic images for model training (Wu et al. 2019). With synthetic data, we can overcome the common class imbalance problem that comes with real-world data.

If the amount of data is still too low for deep learning, we can perform data augmentation with horizontal flips and rotations of the data (Alaudah et al. 2019; Wu et al. 2019). Apart from flip and rotation, we can also use other image transformational techniques, such as shifting, exposure adjustment, contrast change, etc. However, the class imbalance problem may persist after these operations. Xie and Tu (2015) proposed applying a balanced cross-entropy function in such cases, which Wu et al. (2019) implemented in seismic data for structural analysis in the Netherlands, Costa Rica, and Brazil. I recommend applying different strategies, depending on dataset, time, cost, and—most importantly—the underlying geology. Overly sophisticated ML models that cannot resolve real-world geologic problems are not useful and should be discarded.

## 4.7   Recurrent Neural Network and Long Short-Term Memory

We often work with time series data or spatiotemporal data when the occurrence of a previous event influences a current event. These types of problems are common in geosciences, for example when we analyze geochemical, well log, seismic, and production data. These datasets may contain both regional and local patterns. The local patterns in the previous event often influence the current event. Traditional neural networks are not known for using reasoning about previous events to inform later events. It is also due to our assumption that the datasets in ML are independent and identically distributed through their length. Unfortunately, this is not true in most cases, especially with dynamic data (Fig. 4.28). Recurrent neural networks

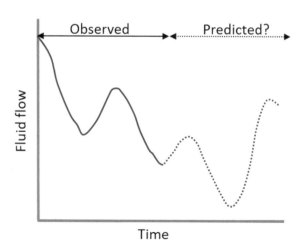

**Fig. 4.28**  An example of water production over time, in which we have a limited number of observations but would like to predict behavior over an extended duration. In theory, these problems are better solved by recurrent neural networks

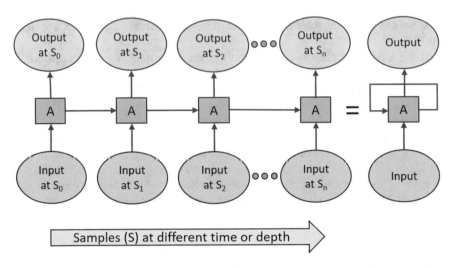

**Fig. 4.29** The simplified architecture of a recurrent neural network. We can use it in sequence data (time series and depth series)

(RNNs) are designed to handle this issue. RNNs can capture the temporal dynamics of sequences. Unlike traditional ANN, the output of RNN not only depends on the current input, but also the previous inputs, making it perfect for sequence analysis. The sequence could be time series, depth series, and text, etc.

RNNs have loops that pass information from one step to the next (Fig. 4.29). We can decompose a RNN into several small, time-dependent neural networks in which the information from a previous network is fed to the next network. We also call these long-term dependencies. Hopfield (1982) introduced the early version of RNNs, which was later modified significantly. Although in theory, RNN can handle long-term dependencies, in reality, it cannot. RNNs can efficiently learn data patterns if the distance between the previous event (from where the relevant information is needed) and the current event is small. If the distance grows, the power of RNNs diminish (vanishing gradient problem). Long short-term memory (LSTM) networks are specially designed to handle this issue.

Hochreiter and Schmidhuber proposed LSTM in 1997. LSTM networks address the vanishing gradient problem found in conventional RNNs by incorporating several gating functions into their state dynamics (Karim et al. 2018). At each time step, an LSTM network contains a hidden vector and a memory vector responsible for controlling state updates and outputs. LSTM networks learn relationships among data during a long-time interval using memory cells that record their states. An LSTM network contains a cell state and three gates (input, output, and forget gates). Figure 4.30 shows an LSTM architecture. The cell state runs straight down the entire chain in the LSTM network, with only some interactions with the gates. This allows the information to transfer along it. Gates regulate the addition or removal of information to or from the network. The forget gate layer controls which value

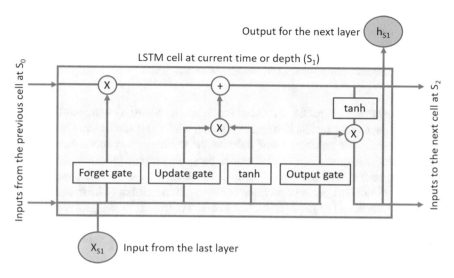

**Fig. 4.30** The simplified architecture of a long short-term memory network

to keep and throw away, expressed in terms of one and zero. The sigmoid function in the forget gate layer passes the information from the previous hidden state and information from the current input.

Why do we need a forget gate layer? It is because we are often interested in certain parts of the sequence, not the other parts, which we want to throw away. The input gate layer decides which information is updated and stored in the cell state. The output gate layer controls what parts of the cell state should propagate to the next layer or output. It basically controls the output flow.

There are several variants of LSTM networks. These include bidirectional LSTM (Schuster and Paliwal 1997), peephole connections (Gers et al. 2000), LSTM with attention (Bahdanau et al. 2014), and multiplicative LSTM (Krause et al. 2016), etc. Each of these variants has certain advantages over others. There have been a limited number of comparative studies of these variants (Greff et al. 2017).

LSTM networks have a few important hyperparameters, such as the number of nodes (neurons), epochs, batch size, layers, dropout rate, regularization, and activation function, etc., which we need to optimize. The number of neurons is an important hyperparameter. Increasing it helps the model performance; however, we run the risk of overfitting. Several diagnostic tests reveal that the number of epochs has control over the LSTM model performance. In general, the error reduces with the increase in the number of epochs; however, after crossing a threshold, the error starts increasing due to overtraining. The batch size controls the frequency of weight updates for the LSTM network. It is the number of samples to be run by the model at a time. In general, the batch size is less than the total number of samples. If the batch size is significantly less than the total number of samples in the dataset, the LSTM model runs fast with less memory requirements, but the model may not be very stable

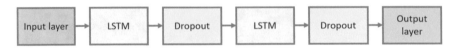

**Fig. 4.31** A schematic diagram of a LSTM with two dropout layers

because the network cannot see most of the data at a time to recognize full patterns in the sequence. This results in higher variance, which is undesirable. On the other hand, increasing the batch size will stabilize the model with reduced variance, but the computational cost will be high. In small datasets, keeping the batch size as close as possible to the total number of samples is ideal. Similar to CNN, we can add dropout to the LSTM network architecture to avoid overfitting (Cheng et al. 2017). Figure 4.31 shows an example of such network. This architecture ignores randomly selected neurons during training and thereby reduces the corresponding weights of those neurons. Thus, the generalization capability of the LSTM network increases with the addition of a dropout layer.

## 4.8  Ensemble Approach

So far, I have discussed several ML algorithms. In general, these algorithms are applied individually to a dataset. Researchers have also studied some of their comparative performances on different types of subsurface data (Bhattacharya et al. 2016, 2019; Di et al. 2018; Zhao 2018). Because each algorithm has its own advantages and disadvantages due to its unique architecture, we cannot really utilize ML's full power unless we combine these algorithms. This is called the ensemble approach, and it has proven successful at improving model performance (Asoodeh et al. 2014; Gholami and Ansari 2017). Ensemble models are meta-algorithms because they combine several individuals to provide the final and best possible solution. Figure 4.32 shows the concept of the ensemble approach. The individual models must be accurate and diverse for an ensemble model to be more accurate than any of its individual members (Hansen and Salamon 1990).

It is often not possible to construct the best possible ensemble models. According to Dietterich (2000), there are three fundamental reasons for this: statistical, computational, and representational. The statistical problem arises when the amount of training data available to the model is significantly small. Without sufficient data, the learning algorithm can find multiple solutions with similar performance. The ensemble model counts the votes from individual models, averages them, and reduces the risk of selecting a poorly performing model (Dietterich 2000). Computational problems are algorithm-specific. Certain algorithms, such as ANN, are prone to get stuck in local minima, even when there are sufficient data. In such cases, we can construct an ensemble of various algorithms to better approximate the true unknown function by running local searches from different start points than the individual models (Dietterich 2000). The third problem, representational, is subtle in nature. In

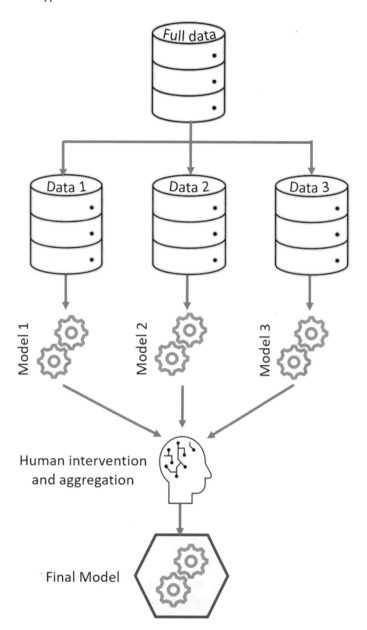

**Fig. 4.32**   The concept of ensemble modeling

many cases, certain algorithms cannot find the true function for classification. An ensemble can form weighted sums of contributions from individual algorithms to expand the space, finding the most representative functions.

There are a few methods to construct an ensemble or committee machine (CM). Gholami and Ansari (2017) combined an optimized neural network, support vector regressor, and fuzzy logic to estimate porosity from seismic attributes. The final output of the CM estimated from the optimized individual models can be mathematically expressed as.

$$Output = w_1 \times ML_1 + w_2 \times ML_2 + \cdots + w_n \times ML_n \qquad (4.4)$$

The weights represent individual contributions of the algorithms, which is an optimization problem.

We can use a bat-inspired algorithm (BA) for optimization (Yang 2010). This algorithm is based on the echolocation behavior bats use to search for prey (Gholami and Ansari 2017; Shiroodi et al. 2017). Bats emit sounds of varying frequency and loudness and they adjust the rate of the emission depending on the proximity to the prey. As the bats move closer to the prey, the sound's loudness decreases and pulse emission rate increases. In the BA algorithm, bat positions are candidate solutions of the problem, and frequency is used to update velocities and, consequently, bats' positions with respect to the prey (Naderi and Khamehchi 2017). In addition, pulse rates and loudness are used to generate a local solution around the selected best global solution. The process is iterated several times until the model reaches convergence.

## 4.9  Physics-Informed Machine Learning

With exponential advances in computational algorithms, resources, and access to data and open-source codes, data-driven modeling has become very popular across scientific disciplines. Although this democratization of AI is an important milestone, it has certain limitations which we need to address now.

Most currently used ML tools, including deep learning, are unable to provide causal insights into the data. Bayesian network-based algorithms can provide a causality direction, depending on the data, but we cannot explain them without domain expertise. It is also well-known that subsurface sampling is highly biased. Based on our geologic knowledge and field analogs, we drill wells and acquire certain data types in certain areas (e.g., 3D seismic, advanced well logs, fiber-optic, core plugs, and fluid samples, etc.). Such data acquisition is time-consuming and expensive, which forces us to work in "small" data capsules that lack diversity. In such cases, the ML network's field of view (FOV) does not cover the full variability possibly present in the system. This profoundly impacts understanding of the physical and chemical processes that form rocks and develop the properties that we measure with surface and downhole sensors. For example, the fluid flow regime changes from transient to boundary dominated over time. If we collect data in only one regime in such a dynamic system, we would not be able to correctly predict fluid flow. In a small data regime, the vast majority of state-of-the-art ML techniques lack robustness and fail to converge (Raissi et al. 2019). We also have no mechanism for comprehending and understanding the meaning of the feature space and attributes computed automatically by deep learning algorithms.

The bottom line is the application of ML, however computationally powerful, will generate serious obstacles and negative impacts to scientific development if we do not integrate them with physical, chemical, biological, and engineering understanding of the system. This is the one of the profound ways we can increase the FOV of ML models in geosciences. We need physics-inspired or physics-informed ML approaches to solve complex geoscience problems. Figure 4.33 shows the concept of physics-informed ML. We can also call it a chemistry-infomed ML if we infuse chemistry-based rules into the model.

The main tenet of ML is that it can map complex non-linear functions with high accuracy in a limited time using limited input features (Raissi et al. 2019). This fundamental proposition fails when the input features are not physics-based or at least physics-aware, when inputs are collected only from certain parts of a system, or when models do not provide causal insights. We should keep in mind that ML algorithms have good interpolation capabilities, but not extrapolation. ML-based systems will not develop the extrapolation ability without physics-based information, at least not in geoscience. There is a well-established body of subsurface geoscience that is based on specific physical/chemical laws and empirical relations based on numerous lab and field-based experiments, such as plate tectonics, sequence stratigraphy, seismic wave propagation, anisotropy, etc. We should use this prior information as an agent of regularization in ML-based modeling, which would allow only a certain set of meaningful solutions and discard other, non-realistic solutions. Moseley et al. (2019)

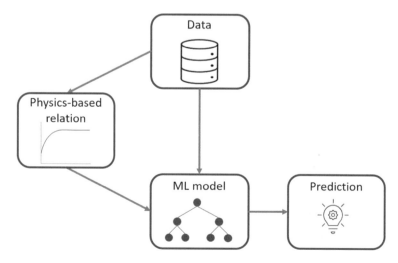

**Fig. 4.33** The concept of physics-informed ML

showed how the addition of causality and physical phenomenon (such as dilation) in deep learning models improved performance in simulating seismic wave propagation and convergence of the solution over the traditional application of CNN algorithms. Physics-informed ML is a growing area of research (Willard et al. 2020) with game-changing potential in geosciences.

# References

Al-Anazi AF, Gates ID (2010) Support vector regression for porosity prediction in a heterogeneous reservoir: a comparative study. Comput Geosci 36(12):1494–1503. https://doi.org/10.1016/j.cageo.2010.03.022

Alaudah Y, Michalowicz P, Alfarraj M, AlRegib G (2019) A machine learning benchmark for facies classification. Interpretation 7(3):SE175–SE187. https://doi.org/10.1190/INT-2018-0249.1

Al-Mudhafar WJM, Bondarenko MA (2015) Integrating K-means clustering analysis and generalized additive model for efficient reservoir characterization. EAGE conference and exhibition

Andoine R, Florea A-C (2020) Weighted random search for CNN hyperparameter optimization. Int J Comput Commun & Control 15(2):3868. https://doi.org/10.15837/ijccc.2020.2.3868

Arthur D, Vassilvitskii S (2007) k-means++: the advantages of careful seeding. Proceedings of the 18th annual ACM-SIAM symposium on discrete algorithms, pp 1027–1035

Asoodeh M, Gholami A, Bagheripour P (2014) Oil-$CO_2$ MMP determination in competition of neural network, support vector regression, and committee machine. J Dispersion Sci Technol 35:564–571. https://doi.org/10.1080/01932691.2013.803255

Badrinarayanan V, Kendall A, Cipolla R (2015) SegNet: a deep convolutional encoder-decoder architecture for image segmentation. IEEE Trans Pattern Anal Mach Intell 39(12):2481–2495. https://doi.org/10.1109/TPAMI.2016.2644615

Bahdanau D, Cho K, Bengio Y (2014) Neural machine translation by jointly learning to align and translate. https://arxiv.org/abs/1409.0473

Ben-Gal I (2007) Bayesian networks. In: Ruggeri F, Faltin F, Kennett R (eds.) Encyclopedia of statistics in quality & reliability. Wiley & Sons. https://doi.org/10.1002/9780470061572.eqr089

Bhattacharya S, Carr TR, Pal M (2016) Comparison of supervised and unsupervised approaches for mudstone lithofacies classification: case studies from the Bakken and Mahantango-Marcellus Shale, USA. J Nat Gas Sci Eng 33:1119–1133. https://doi.org/10.1016/j.jngse.2016.04.055

Bhattacharya S, Ghahfarokhi PK, Carr TR, Pantaleone S (2019) Application of predictive data analytics to model daily hydrocarbon production using petrophysical, geomechanical, fiber-optic, completions, and surface data: a case study from the Marcellus Shale, North America. J Petrol Sci Eng 176:702–715. https://doi.org/10.1016/j.petrol.2019.01.013

Bhattacharya S, Mishra S (2018) Applications of machine learning for facies and fracture prediction using Bayesian Network Theory and Random Forest: case studies from the Appalachian basin, USA. J Petrol Sci Eng 170:1005–1017. https://doi.org/10.1016/j.petrol.2018.06.075

Bishop C (1995) Pattern recognition and machine learning. Springer

Bouckaert RR (1995) Bayesian belief networks: from construction to inference. PhD thesis. University of Utrecht

Bouckaert RR (2008) Bayesian network classifiers in weka for version 3-5-7. https://www.cs.waikato.ac.nz/~remco/weka.bn.pdf

Breiman L (2001) Random forests. Mach Learn 45:5–32. https://doi.org/10.1023/A:1010933404324

Carvalho AM (2009) Scoring functions for learning bayesian networks. INESC-ID technical report 54/2009

Celebi ME, Kingravi HA, Vela PA (2012) A comparative study of efficient initialization methods for the K-Means clustering algorithm. https://arxiv.org/pdf/1209.1960.pdf

Cheng G, Peddinti V, Povey D, Manohar V, Khudanpur S, UYan Y (2017) An exploration of dropout with LSTMs. https://www.danielpovey.com/files/2017_interspeech_dropout.pdf

Christianini N, Shawe-Taylor J (2000) An introduction to support vector machines and other kernel-based learning methods. Cambridge University Press. https://doi.org/10.1017/CBO978051180 1389

Coléou T, Poupon M, Azbel K (2003) Unsupervised seismic facies classification: A review and comparison of techniques and implementation. Lead Edge 22(10):942–953. https://doi.org/10. 1190/1.1623635

Cooper GF, Herskovits E (1992) A Bayesian method for the induction of probabilistic networks from data. Mach Learn 9:309–347. https://doi.org/10.1007/BF00994110

Cortes C, Vapnik V (1995) Support-vector networks. Mach Learn 20:273–297. https://doi.org/10. 1007/BF00994018

Cutler A, Cutler DA, Stevens JR (2011) Random forests. Mach Learn 1–19

Di H, Li Z, Maniar H, Abubakar A (2019) Seismic stratigraphy interpretation via deep convolutional neural networks. SEG Technical Program Expanded Abstracts, 2358–2362. https://doi.org/10. 1190/segam2019-3214745.1

Di H, Wang Z, AlRegib G (2018) Seismic fault detection from post-stack amplitude by convolutional neural networks. Conference proceedings, 80th EAGE conference and exhibition, pp 1–5. https:// doi.org/10.3997/2214-4609.201800733

Dietterich TG (2000) Ensemble methods in machine learning. Lecture Notes in Computer Science 1857. Springer, Berlin, Heidelberg. https://doi.org/10.1007/3-540-45014-9_1

Ding G, Qin L (2020) Study on the prediction of stock price based on the associated network model of LSTM. Int J Mach Learn Cybern 11:1307–1317. https://doi.org/10.1007/s13042-019-01041-1

Doveton JH (1994) Geologic log analysis using computer methods. American association of petroleum geologists

Dramsch JS, Lüthje M (2018) Deep-learning seismic facies on state-of-the-art CNN architectures. SEG Technical Program Expanded Abstracts, 2036–2040. https://doi.org/10.1190/segam2018-2996783.1

Forgy E (1965) Cluster analysis of multivariate data: efficiency vs interpretability of classification. Biometrics 21:768

Gers FA, Schmidhuber J, Cummins F (2000) Learning to forget: continual prediction with LSTM. Neural Comput 12(10):2451–2471. https://doi.org/10.1162/089976600300015015

Gholami A, Ansari HR (2017) Estimation of porosity from seismic attributes using a committee model with bat-inspired optimization algorithm. J Petrol Sci Eng 152:238–249. https://doi.org/ 10.1016/j.petrol.2017.03.013

Greff K, Srivastava RK, Koutník J, Steunebrink BR, Schmidhuber J (2017) LSTM: a search space odyssey. Trans Neural Netw Learn Syst 28(10):2222–2232. https://doi.org/10.1109/TNNLS. 2016.2582924

Hansen LK, Salamon P (1990) Neural network ensembles. IEEE Trans Pattern Anal Mach Intell 12(10):993–1001. https://doi.org/10.1109/34.58871

Hastie T, Tibshirani R (1998) Classification by pairwise coupling. Ann Stat 26(2):451–471. https:// doi.org/10.1214/aos/1028144844

He K, Zhang X, Ren S, Sun J (2016) Deep residual learning for image recognition. IEEE conference on computer vision and pattern recognition, Las Vegas, NV, pp 770–778. https://doi.org/10.1109/ CVPR.2016.90

Heckerman D, Geiger D, Chickering DM (1995) Learning Bayesian networks: the combination of knowledge and statistical data. Mach Learn 20:197–243. https://doi.org/10.1023/A:102262321 0503

Hernán MA, Robins JM (2006) Instruments for causal inference: an epidemiologist's dream? Epidemiology 17(4):360–372. https://doi.org/10.1097/01.ede.0000222409.00878.37

Hinz T, Navarro-Guerrero N, Magg S, Wermter S (2018) Speeding up the hyperparameter optimization of deep convolutional neural networks. Int J Comput Intell Appl 17(2):1850008. https://doi. org/10.1142/S1469026818500086

Hochreiter S, Schmidhuber J (1997) Long short-term memory. Neural Comput 9(8):1735–1780. https://doi.org/10.1162/neco.1997.9.8.1735

Holdaway KR, Irving DHB (2017) Enhance oil & gas exploration with data-driven geophysical and petrophysical models. Wiley

Hopfield JJ (1982) Neural networks and physical systems with emergent collective computational abilities. PNAS 79(8):2554–2558

Huang J, Yuan Y, Cui W, Zhan Y (2008) Development of data mining application for agriculture based Bayesian networks. In: IFIP international federation for information processing, 258: Computer and computing technologies in agriculture, vol 1. Springer, Li Daoling, Boston, pp 645–652

Huang L, Dong X, Clee TE (2017) A scalable deep learning platform for identifying geologic features from seismic attributes. Lead Edge 36(3):249–256. https://doi.org/10.1190/tle360302 49.1

Karim F, Majumdar S, Darabi H, Chen S (2018) LSTM fully convolutional networks for time series classification. IEEE Access 6:1662–1669. https://doi.org/10.1109/ACCESS.2017.2779939

Kordon A (2010) Applying computational intelligence. Springer

Krause B, Lu L, Murray I, Renals S (2016) Multiplicative LSTM for sequence modelling. ICLR. arXiv:1609.07959

Kuzma HA (2003) A support vector machine for AVO interpretation. SEG Technical Program Expanded Abstracts, 181–184

Liu CHB, Chamberlain BP, Little DA, Cardoso AM (2017) Generalising random forest parameter optimisation to include stability and cost. https://arxiv.org/pdf/1706.09865.pdf

Louppe G (2014) Understanding random forests: from theory to practice. PhD dissertation, University of Liège. https://arxiv.org/pdf/1407.7502.pdf

Loussaief S, Abdelkrim A (2018) Convolutional neural network hyper-parameters optimization based on genetic algorithms. Int J Adv Comput Sci Appl 9(10):252–266. https://doi.org/10. 14569/IJACSA.2018.091031

Luts J, Ojeda F, Van de Plas R, De Moor B, Van Huffel S, Suykenset JAK (2010) A tutorial on support vector machine-based methods for classification problems in chemometrics. Anal Chim Acta 665(2):129–145. https://doi.org/10.1016/j.aca.2010.03.030

MacQueen J (1967) Some methods for classification and analysis of multivariate observations. Proceedings of the fifth Berkeley symposium on mathematical statistics and probability, Vol 1, pp 281–297

Matos MC, Osorio PLM, Johann PRS (2007) Unsupervised seismic facies analysis using wavelet transform and self-organizing maps. Geophysics 72(1):9–21

McCulloch WS, Pitts W (1943) A logical calculus of the ideas immanent in nervous activity. Bull Math Biophys 5:115–133

Mishra S, Datta-Gupta A (2018) Applied statistical modeling and data analytics. Elsevier

Misra S, Li H, He J (2019) Machine learning for subsurface characterization. Gulf Professional Publishing

Mboga N, Persello C, Bergado JR, Stein A (2017) Detection of informal settlements from vhr images using convolutional neural networks. Remote Sens 9(11):1106. https://doi.org/10.3390/ rs9111106

Mohaghegh SD (2017) Shale analytics. Springer

Moseley B, Nissen-Meyer T, Markham A. (2019) Deep learning for fast simulation of seismic waves in complex media. Solid Earth Discussions. https://doi.org/10.5194/se-2019-157

Naderi M, Khamehchi E (2017) Well placement optimization using metaheuristic bat algorithm. J Petrol Sci Eng 150:348–354. https://doi.org/10.1016/j.petrol.2016.12.028

Pal M, Foody GM (2010) Feature selection for classification of hyperspectral data by SVM. IEEE Trans Geosci Remote Sens 48(5):2297–2307

Pal M, Foody GM (2012) Evaluation of SVM, RVM and SMLR for accurate image classification with limited ground data. IEEE J Sel Top Appl Earth Obs Remote Sens 5(5):1344–1355. https:// doi.org/10.1109/JSTARS.2012.2215310

Pearl J (2009) Causality, 2nd edn. Cambridge University Press

Pearl J, Verma TS (1991) A theory of inferred causation. https://citeseerx.ist.psu.edu/

Pham N, Fomel S, Dunlap D (2019) Automatic channel detection using deep learning. Interpretation 7(3):SE43–SE50. https://doi.org/10.1190/INT-2018-0202.1

Raissi M, Perdikaris P, Karniadakis GE (2019) Physics-informed neural networks: a deep learning framework for solving forward and inverse problems involving nonlinear partial differential equations. J Comput Phys 378:686–707. https://doi.org/10.1016/j.jcp.2018.10.045

Russakovsky O, Deng J, Su H, Krause J, Satheesh S, Ma S, Huang Z, Karpathy A, Khosla A, Bernstein M, Berg AC, Fei-Fei L (2015) ImageNet large scale visual recognition challenge. Int J Comput Vision 115(3):211–252. https://doi.org/10.1007/s11263-015-0816-y

Schuster M, Paliwal KK (1997) Bidirectional recurrent neural networks. IEEE Trans Signal Process 45(11):2673–2681

Scutari M, Vitolo C, Tucker A (2019) Learning Bayesian networks from big data with greedy search: Computational complexity and efficient implementation. Stat Comput 29:1095–1108. https://doi.org/10.1007/s11222-019-09857-1

Sen S, Kainkaryam S, Ong C, Sharma A (2019) Regularization strategies for deep-learning-based salt model building. Interpretation 7(4):T911–T922. https://doi.org/10.1190/INT-2018-0229.1

Shiroodi SK, Ghafoori M, Ansari HR, Lashkaripour G, Ghanadian M (2017) Shear wave prediction using committee fuzzy model constrained by lithofacies, Zagros basin, SW Iran. J Afr Earth Sc 126:123–135. https://doi.org/10.1016/j.jafrearsci.2016.11.016

Smith LN (2018) A disciplined approach to neural network hyper-parameters: Part 1—learning rate, batch size, momentum, and weight decay. https://arxiv.org/pdf/1803.09820.pdf

Steinhaus H (1956) Sur la division des corps matériels en parties. Bulletin De l'Académie Polonaise Des Sciences, Classe III, IV(12):801–804

Su T, Dy JG (2007) In search of deterministic methods for initializing K-Means and Gaussian mixture clustering. Intell Data Anal 11(4):319–338

Thornley S, Marshall RJ, Wells S, Jackson R (2013) Using directed acyclic graphs for investigating causal paths for cardiovascular disease. J Biom & Biostat 4(5). https://doi.org/10.4172/2155-6180.1000182

Wang G, Carr TR, Ju Y, Li C (2014) Identifying organic-rich Marcellus Shale lithofacies by support vector machine classifier in the Appalachian basin. Comput Geosci 64:52–60. https://doi.org/10.1016/j.cageo.2013.12.002

Willard J, Jia X, Xu S, Steinbach M, Kumar V (2020) Integrating physics-based modeling with machine learning: a survey. https://arxiv.org/pdf/2003.04919.pdf

Wu X, Liang L, Shi Y, Fomel S (2019) FaultSeg3D: using synthetic datasets to train an end-to-end convolutional neural network for 3D seismic fault segmentation. Geophysics 84(3):IM35–IM45. https://doi.org/10.1190/geo2018-0646.1

Xie S, Tu Z (2015) Holistically-nested edge detection. Proceedings of the IEEE international conference on computer vision, pp 1395–1403. https://doi.org/10.1109/ICCV.2015.164.

Yang XS (2010) A new metaheuristic bat-inspired algorithm, nature inspired cooperative strategies for optimization. Stud Comput Intell 284:65e74

Zhang H, Liu Y, Zhang Y, Xue H (2019) Automatic seismic facies interpretation based on an enhanced encoder-decoder structure. SEG Technical Program Expanded Abstracts, 2408–2412. https://doi.org/10.1190/segam2019-3215516.1

Zhao T (2018) Seismic facies classification using different deep convolutional neural networks. SEG Technical Program Expanded Abstracts, 2046–2050. https://doi.org/10.1190/segam2018-2997085.1

Zhong Z, Carr TR (2016) Application of mixed kernels function (MKF) based support vector regression model (SVR) for $CO_2$—Reservoir oil minimum miscibility pressure prediction. Fuel 184:590–603. https://doi.org/10.1016/j.fuel.2016.07.030

# Chapter 5
# Summarized Applications of Machine Learning in Subsurface Geosciences

**Abstract** Geoscientists have been implementing machine learning (ML) algorithms for several classifications and regression related problems in the last few decades. ML's implementation in geosciences came in different phases, and often these broadly followed or lagged after certain advances in computer sciences. We can trace back some of the early applications of modern ML techniques to 1980–1990. Geoscientists were mostly dealing with deterministic analytical solutions at that time, and they were encouraged to do so at their organizations. This is also the time when geostatistics started flourishing in reservoir characterization and modeling efforts. Then, the early 2000's saw a slight uptick in ML applications, mostly neural networks and decision trees. Since 2014–2015, a lot of ML-related work was published. In addition to open-source languages, this also has to do with access to the massive volume of data from unconventional reservoirs. And then, since 2017, there has been an explosion of deep learning related work. This again corresponds to the convolutional neural network architecture published by Goodfellow in 2014. Initially, the ML work in geosciences focused on petrophysics, seismic, and now core and thin section images. Another growing trend is the application of ML in passive geophysical data analysis (seismology, gravity, and magnetic, etc.). As of now, most of the published studies on ML are confined to outlier detection, facies, fracture, and fault classification, rock property (e.g., poro-perm-fluid saturation-total organic carbon-geomechanics) prediction, predicting missing logs/variables, and well log correlation. In this chapter, we will review some of these popular research problems tackled by ML.

**Keywords** Outlier detection · Facies classification · Fault classification · Reservoir property prediction · Geomechanical property prediction · Fluid flow prediction

## 5.1 Outlier Detection

Access to high-quality data is the foremost issue for building ML-based models. Often, our subsurface datasets are noisy and contain outliers. Outliers skew the statistical relations among parameters. Bad data points affect any inversion modeling and ML-based classification and regression results. If we use such data in our modeling

© The Author(s), under exclusive license to Springer Nature Switzerland AG 2021     123
S. Bhattacharya, *A Primer on Machine Learning in Subsurface Geosciences*,
SpringerBriefs in Petroleum Geoscience & Engineering,
https://doi.org/10.1007/978-3-030-71768-1_5

(e.g., facies classification), the results will not be meaningful and consistent. It will result in wrong estimates of rock and fluid properties, and ultimately reservoir estimates. Therefore, quality assurance and check (QA/QC) of data are important. Outliers can result from different reasons, including subsurface conditions (e.g., borehole washout), tool functions, and the formation itself. For the last one, we need to be aware of the geologic context. We can remove, replace, and transform outliers, depending on the nature of the outlier, dataset, and problem, but we need to detect them first. The problem is implementing a consistent outlier detection process in a large database covering a large area. For example, density and neutron logging tools get affected by bad boreholes, measured by the caliper log. Often, there are washout zones in the borehole. Traditionally, petrophysicists look at the individual wells and flag the bad measurement zones based on the caliper log response and reconstruct the logs manually to the best possible extent (Fig. 5.1). In addition, bad boreholes (e.g., large, rugose, and tight) affect different wireline logs differently, which makes the process more difficult. This takes a significant amount of time, and sometimes the results may not be consistent across the whole basin, depending on the tool types, variable responses, vendors, and calibration. We can use ML to identify clusters of

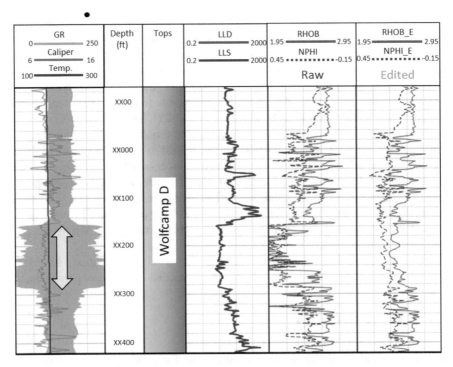

**Fig. 5.1** An example of a borehole washout zone (yellow arrow), corresponding responses of bulk density and neutron porosity logs (raw), and manually edited density and neutron porosity logs

good and bad measurements in an unsupervised manner. Sen et al. (2020) demonstrated an example of the ML-assisted automatic detection of bad density measurement in hundreds of wells in the Permian Basin in the United States (Fig. 5.2). They used an unsupervised time-series clustering algorithm (Toeplitz Inverse Covariance Clustering, TICC) on the caliper and density logs to automatically generate labels of good and bad log response. Once they generated the labels, they used supervised ML algorithms to predict the bad measurements in boreholes, with no available caliper logs. They used the synthetic minority oversampling technique (SMOTE) technique to balance the samples corresponding to good and bad data because the number of outliers is always smaller than the actual signal; otherwise, ML-based results would have been highly skewed. Misra et al. (2019) used unsupervised ML algorithms, such as one-class support vector machine (SVM) and density-based spatial clustering of applications with noise (DBSCAN), to detect outliers in well log data. Such applications of class-based ML are beneficial in geosciences.

## 5.2 Petrophysical Log Analysis

### 5.2.1 Facies Clustering and Classification

Petrophysical log-based facies (also referred to as electrofacies or petrofacies) clustering and classification is a trendy area of research for ML applications. Geoscientists have been working on this particular problem for the last two decades. Facies control petrophysical properties and fluid flow. Accurate identification and prediction of facies are critical to infer the multi-scale depositional/diagenetic trends and build meaningful and manageable reservoir models. It is not possible to acquire core samples in all wells due to economic and logistic constraints. ML has been very useful in solving these problems with demonstrated success.

### 5.2.2 General Rationale behind the use of Conventional Well Logs for Facies Identification

Gamma-ray: Mudstones show a higher gamma-ray response due to high uranium content than sandstones and limestones.
Deep resistivity: Organic matter has high resistivity, therefore, it is useful to identify organic-rich mudstones.
Bulk density: Different rock-forming minerals have different densities (quartz: 2.65 g/cc, calcite: 2.71 g/cc, and dolomite: 2.87 g/cc). Organic-rich mudstones have low density due to the presence of kerogen (density: 1–1.4 g/cc).
Neutron porosity: Mudstones show high neutron porosity than sandstones and carbonates due to clay-bound water.

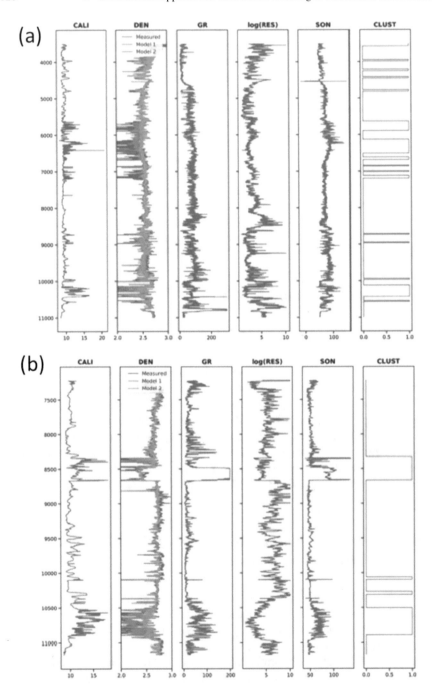

**Fig. 5.2** Density log prediction results for two test wells **a** and **b** using light gradient boosting machine-based models trained on a large dataset in the Permian Basin (United States) before (Model 1) and after (Model 2) removal of bad hole sections, as predicted by TICC algorithm (Sen et al. 2020) (Permission granted from SPWLA)

Photoelectric: Carbonate rocks have higher photoelectric (PE) responses compared to clastics (sandstone and shale). Calcite and dolomite have characteristic PE responses of ~5 b/e and ~3 b/e, respectively. Mudstones can be carbonate-rich too. Calcareous mudstones (e.g., Eagle Ford Shale) and carbonate interlayers inside the mudstone successions (e.g., Marcellus and Wolfcamp Shale) can be easily distinguished from the siliceous and clay-rich mudstones due to their high photoelectric values.

Apart from the above well logs, we can generate new predictors, such as effective porosity, total organic carbon, RHOmaa, and Umaa.

### 5.2.3  Machine Learning for Well-Log-Based Facies Clustering and Classification

Qi and Carr (2006) used a single-hidden layer back-propagation artificial neural network (ANN) for facies classification in the Mississippian-age St. Louis limestone reservoir in Kansas, United States. It was deposited across an extensive shallow-marine carbonate shelf with the periodic occurrence of a nonmarine facies (Handford and Francka 1991; Abegg et al. 2001). As expected in such an environment, we can see a multitude of facies variabilities being present. The formation comprises six facies, including quartz-rich carbonate grainstone (lithofacies 1), argillaceous limestone (lithofacies 2), skeletal wackestone (lithofacies 3), skeletal grainstone/packstone (lithofacies 4), porous ooid grainstone (lithofacies 5), and cemented ooid grainstone (lithofacies 6). Qi and Carr (2006) used ten cored wells with conventional logs as predictors (e.g., gamma-ray, medium and deep resistivity, density porosity, neutron porosity, and photoelectric) to train and test the ANN model (Fig. 5.3). Then they used that model to predict facies for 90 uncored wells. They used two-thirds of the wells for training and the remaining one-third for testing the model performance before applying it to the 90 uncored wells. The ultimate product from their work was several cross-sections showing vertical and lateral heterogeneities of carbonate facies used in geomodeling (Fig. 5.4). Their facies classification accuracy varied between 70.37% and 90.82%. One important conclusion from their work was that the addition of adjoining facies within-one lithofacies improved accuracy slightly (93.72%). This is important as often there are specific rules of facies associations, which we need to recognize and incorporate into the ML model during training.

Howat et al. (2016) showed the application of ML in predicting vugs in the Copper Ridge Formation of the Appalachian Basin, United States, related to a carbon sequestration project. Vugs are irregular cavities inside carbonate rocks, formed by dissolution processes that may result in higher permeability zones, and they can store fluid in enough volume. Vugs are hard to be readily identified using conventional well logs. Howat et al. (2016) used high-resolution image logs and core samples from six wells to identify vugs and trained/tested ML models using conventional well logs. They used nine ML algorithms, including SVM, KNN, RF, and Naïve Bayes, etc. They assessed the model performance using well-level cross-validation, which means that

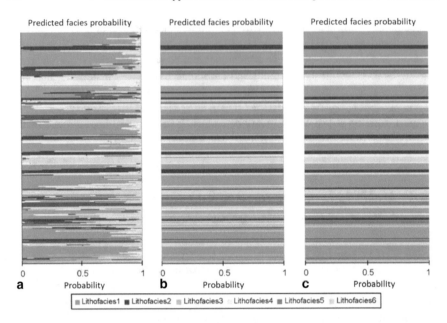

**Fig. 5.3** Predicted lithofacies probability curve, predicted discrete lithofacies curve, and actual lithofacies curve for 10 cored wells, St. Louis Limestone: **a** predicted facies probability plot; **b** predicted lithofacies plot; and **c** actual lithofacies plot (Qi and Carr 2006) (Reprinted from Computers & Geosciences, 32/7, L Qi and TR Carr, Neural network prediction of carbonate lithofacies from well logs, Big Bow and Sand Arroyo Creek fields, Southwest Kansas, 947–964, Copyright (2006), with permission from Elsevier)

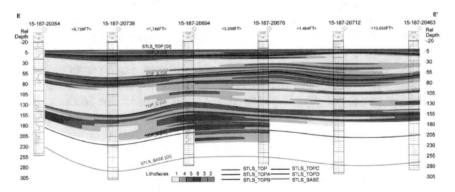

**Fig. 5.4** Interpolated color filled lithofacies cross-section (Qi and Carr 2006) (Reprinted from Computers & Geosciences, 32/7, L Qi and TR Carr, Neural network prediction of carbonate lithofacies from well logs, Big Bow and Sand Arroyo Creek fields, Southwest Kansas, 947–964, Copyright (2006), with permission from Elsevier)

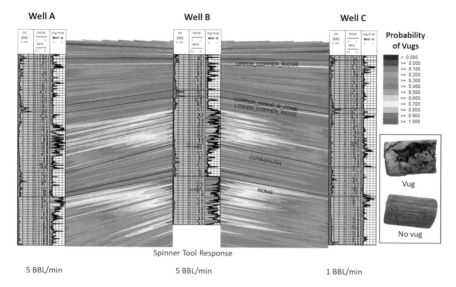

**Fig. 5.5** Vug prediction across a few wells (modified after Howat et al. 2016). Spinner tool response indicates injectivity of fluid. Note wells with higher vug probability have higher injectivity. This is critical to fluid storage (water, carbon, and hydrogen) studies (Permission received)

they trained the model using five wells and tested it on the held-out well in a recursive manner until all wells were tested. They found the SVM model as the top performer with a 78% correct classification rate. Their ultimate product was a vug model with probabilistic estimates (Fig. 5.5). Recently, Deng et al. (2019) successfully implemented used SVM, RF, and ANN to predict vugs using conventional logs, nuclear magnetic resonance, and core data in Kansas. They found the SVM classifier to be the most efficient algorithm.

Mudstone facies classification is considered relatively complex compared to sandstone and carbonate formations for several reasons. Mudstones are heterogeneous at different scales, and they have more vertical heterogeneity than lateral heterogeneity because of depositional conditions. Wang (2012), Bhattacharya et al. (2015), Bhattacharya et al. (2016), and Bhattacharya and Mishra (2018) classified several mudstone facies in the Marcellus and Bakken Shales in North America using ML algorithmss. These are prolific hydrocarbon source-rocks. Using 10 conventional well logs (including feature-engineered petrophysical response), Bhattacharya et al. (2016) and Bhattacharya and Mishra (2018) classified up to six mudstone facies in the Marcellus Shale (Fig. 5.6). They showed that RF and SVM models could classify with up to 81–82% accuracy. They also used Bayesian Network to provide insights into the relationship between predictors and facies. They show that gamma-ray, deep resistivity, and bulk density logs are more influential than others to classify mudstone facies. In addition to classification, they also used multi-resolution graph-based clustering (MRGC) to compare the results with supervised classification. It

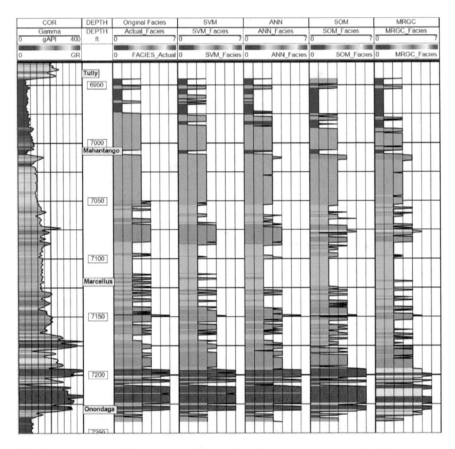

**Fig. 5.6** An example of well log-based carbonate and mudstone facies classification using four ML algorithms (e.g., SVM, ANN, SOM, and MRGC) across the Tully-Marcellus interval in the Appalachian Basin, United States (after Bhattacharya et al. 2016). SVM and ANN are better predictors than SOM and MRGC (Reprinted from Journal of Natural Gas Science and Engineering, 33, S Bhattacharya, TR Carr, and M Pal, Comparison of supervised and unsupervised approaches for mudstone lithofacies classification: Case studies from the Bakken and Mahantango-Marcellus Shale, USA, 1119–1133, Copyright (2016), with permission from Elsevier)

is important to realize that well log-based facies have coarser resolution than core-based facies. Similar studies can be done with core-based continuous rock properties (e.g., core-based spectral gamma-ray and x-ray fluorescence profiles).

Most often, ML-based facies classification studies stop at basic $R^2$ and simple error estimates. Based on several studies, it does appear that we find errors in classification in two different patterns. The first one happens at the boundaries of formation/members, and the second one happens inside the same class when one of the predictors change its behavior drastically from its previous sample in the same well or study area. Many times, the first type of error happens due to the shoulder-bed effect (bed thickness below log resolution) and the inability of ML to extrapolate beyond

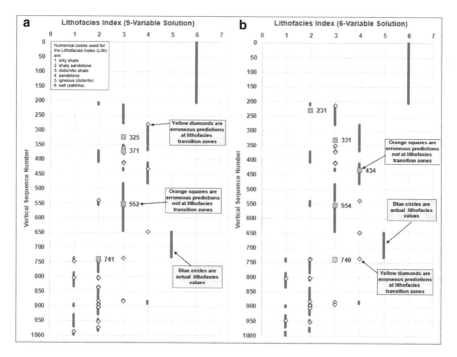

**Fig. 5.7** An example of the application of a transparent open box algorithm for facies classification in a gas field in Algeria. This algorithm attempts to improve facies prediction near transition zones (Wood 2019) (Reprinted from Marine and Petroleum Geology, 110, DA Wood, Lithofacies and stratigraphy prediction methodology exploiting an optimized nearest-neighbour algorithm to mine well-log data, 347–367, Copyright (2019), with permission from Elsevier)

the boundaries. This can particularly affect mudstones and laminated shaly-sand sequences (composed of thin beds). To tackle these challenges, Wood (2018, 2019) proposed the application of a new algorithm- transparent open box (TOB) learning network, which is a modified version of RF. This network rather focuses on reducing errors in classification. Wood (2019) applied this algorithm with a moderate success to the Triassic reservoir section of the giant Hassi R'Mel gas field in Algeria. As per Wood (2019), TOB-based modeling can provide an understanding of the causes of the few data prediction errors, and these errors can be rectified in several instances if properly calibrated (Fig. 5.7).

## 5.2.4  Fracture Classification

Fractures control fluid flow. A detailed understanding of the presence of fractures and their geometry is crucial to producing fluid from the tight formations. However, fractures are hard to be readily identified using conventional well logs. We use core samples, petrographic thin sections, micro-CT images, and image logs for fracture identification. Keep in mind that the access to core samples and image logs is limited due to cost and time, which is why, ML-based fracture classification using conventional logs can be a good alternative approach. In addition, the interpretation of image logs can be highly subjective at times.

## 5.2.5 General Rationale behind the use of Conventional Well Logs for Fracture Classification

Caliper: The caliper log typically shows two types of responses in a fracture zone. It may show borehole elongation along the main direction of fracture orientations due to breakage of fractured rocks during drilling (Tokhmchi et al. 2010). Caliper response may also indicate a reduced borehole size, since high permeability in fracture zones leads to the presence of a thick mud cake, especially when lost circulation material or heavily weighted mud is employed (Tokhmchi et al. 2010). As caliper log alone cannot give precise and readily interpretable information about fractures, a secondary attribute (called Delta_CALI) can be calculated by subtracting the bit size from the caliper response (Bhattacharya and Mishra 2018). A high positive value indicates fractures and/or cavings (weak formation), whereas a high negative value indicates tight spots. Two-arm caliper logs are useful to separate wellbore breakout from washout zones.

Gamma-ray: Sometimes, an increase in gamma-ray is observed in fractured reservoirs without concurrently higher shale volume due to uranium salts' deposition along the fractures. Access to spectral gamma-ray could be a good predictor in such cases.

Sonic: An increase in travel time is expected due to the presence of fractures (if open or fluid-filled).

Bulk Density: We can expect a reduction in density in case of open fractures due to an increase in total porosity.

Resistivity: We often find a difference between shallow and deep resistivity logs in open fractures (Vasvári 2011) and invaded by drilling mud. We have to be careful about the sign of the deviation as it is controlled by the resistivity of the mud filtrate (Shazly and Tarabees 2013).

## 5.2.6 Machine Learning for Well-Log-Based Fracture Classification

There have been limited studies on ML-based fracture classification using well logs (Ja'fari et al. 2011; Zazoun 2013; Bhattacharya and Mishra 2018; Dong et al. 2020). There are a few major issues to fracture identification using well logs; first, no individual conventional wireline logs can identify fractures effectively. Second, it is a problem with an imbalanced dataset (far less fractures than no fractures). Zazoun (2013) studied fractures from core and conventional well logs using ANN in the Cambro-Ordovician sandstone reservoir of Mesdar oil field, Algeria. Zazoun (2013) used core measurements from 13 wells and used it to supervise the ANN model with conventional logging suites (e.g., caliper, gamma-ray, sonic, density, and neutron porosity) (Fig. 5.8). Three types of fractures were identified, including open, sealed, and closed fractures. After applying feature scaling, Zazoun assigned 70% of the

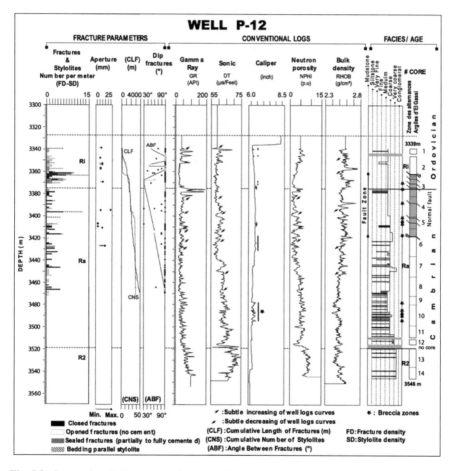

**Fig. 5.8**  Core and equivalent conventional logs showing fracture parameters and grain size distribution in a well in the Saharan Platform in Algeria (Zazoun 2013). GR, DT, Caliper, NPHI, and RHOB logs have a subtle response to fractures. The caliper log shows changes in its response due to the presence of larger fractures or fracture zone. In case of open fractures, density and velocity log responses decrease. Black bars in the first track from the left indicate the fractured zones (Reprinted from Journal of African Earth Sciences, 83, RS Zazoun, Fracture density estimation from core and conventional well logs data using artificial neural networks: The Cambro-Ordovician reservoir of Mesdar oil field, Algeria, 55–73, Copyright (2013), with permission from Elsevier)

training data and the remaining 30% for test and validation. The ANN with the conjugate gradient descent approach performed better than other ANN models (e.g., back-propagation). The model showed a high correlation coefficient ($R^2 = 0.948$) between real fractures and predicted fracture density (Fig. 5.9).

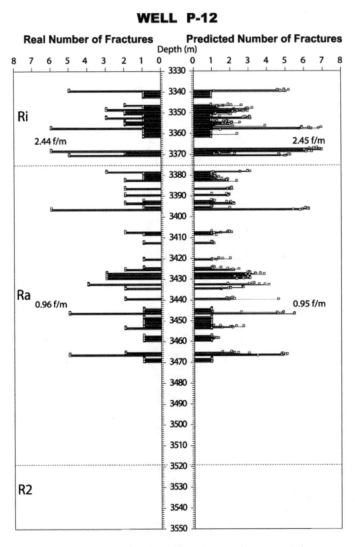

**Fig. 5.9** A comparison between the real and predicted fracture density in the test well in the Saharan Platform in Algeria (Zazoun 2013). The numbers of real and predicted fractures per meter present approximately the same value for the Ri and Ra Units (Reprinted from Journal of African Earth Sciences, 83, RS Zazoun, Fracture density estimation from core and conventional well logs data using artificial neural networks: The Cambro-Ordovician reservoir of Mesdar oil field, Algeria, 55–73, Copyright (2013), with permission from Elsevier)

## 5.2.7  Well-Log-Based Rock Property Prediction

### 5.2.7.1  Porosity, Permeability, and Fluid Saturation Prediction

Poro-perm and fluid saturation are critical petrophysical properties. We can derive porosity from conventional well logs, but we need different empirical equations and advanced logs such as NMR to derive permeability and fluid saturation. We often acquire such data from routine core analyses and calibrate them to wireline log responses. Similar to classification, geoscientists have been using ML for regression-related problems for the last two decades (Bhatt 2002; Al-Anazi and Gates 2010; Zhong et al. 2019).

Helle et al. (2001) and Bhatt (2002) applied ANN to predict porosity and permeability using conventional well logs for a basin-wide fluid flow analysis project in the Viking Graben, North Sea. Bhatt used both synthetic and real well log data (e.g., density, sonic, and resistivity) for model training and implementation (Fig. 5.10). Core-based grain density data were used to derive the porosity from the core samples; these are the best possible estimates of in-situ porosity values. 80% of the data was used for model training and the remaining 20% for testing. They trained 20 neural networks with the same input data but with different initial weights, out of which they selected nine networks with minimum bias using a committee machine approach. Helle et al. (2001) showed the ANN-based model showed R in one well 0.89 (Fig. 5.11). ML-based porosity showed a good match with core-based porosity in most of the formations under study; however, it did not work well in the formations with thin beds and coal layers. This is beacuse input well logs have coarser resolution than thin beds; therefore, the ML-model trained using such input data yielded a low-resolution prediction. In such cases, the application of high-resolution advanced petrophysical logs or even feeding some core-based information (such as empirical relations or transforms) back to the ML-model could be useful.

Since the 2000s, geoscientists used several traditional ML algorithms to predict poro-perm values successfully (Mohaghegh and Ameri 1995; Rogers et al. 1995; Bhatt 2002; Al-Anazi and Gates 2012). Bhatt and Helle (2002) and Bhatt (2002) also used ML to predict permeability using conventional well logs. In general, we estimate permeability using the Kozeny-Carman equation, NMR logs, and well test data. The common practice is using core-based porosity–permeability transforms and using that to predict permeability from porosity logs. Bhatt and Helle (2002) implemented a simple neural network and a modular neural network on both synthetic and real log data (e.g., gamma-ray, density, neutron, and sonic) to predict permeability. Similar to porosity prediction, they used synthetic log data to design the optimal ML-architecture. Because the permeability values have a large range, the training dataset was split into three permeability ranges, and then both ensemble and modular combinations were applied on. Each module was assigned to predict permeability in a given range, and the modules, in turn, are combined to cover the entire range. Figure 5.12 shows the model-driven results, bias, and variance. The final R values after ML-based modeling varied between 0.73 and 0.83 (Bhatt 2002).

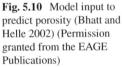

**Fig. 5.10** Model input to predict porosity (Bhatt and Helle 2002) (Permission granted from the EAGE Publications)

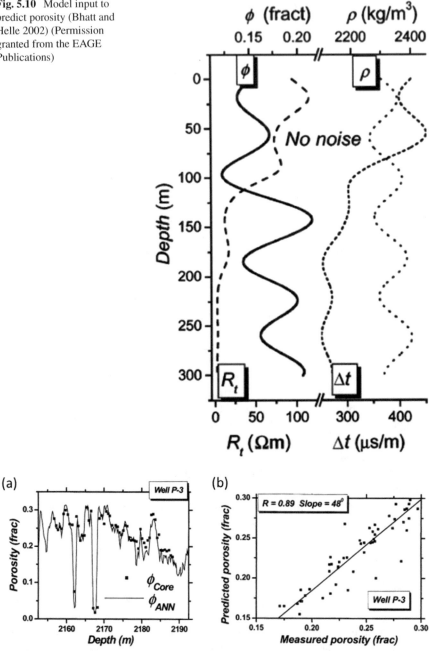

**Fig. 5.11** Model performance on predicting porosity (Helle et al. 2001) (Permission granted from the EAGE Publications)

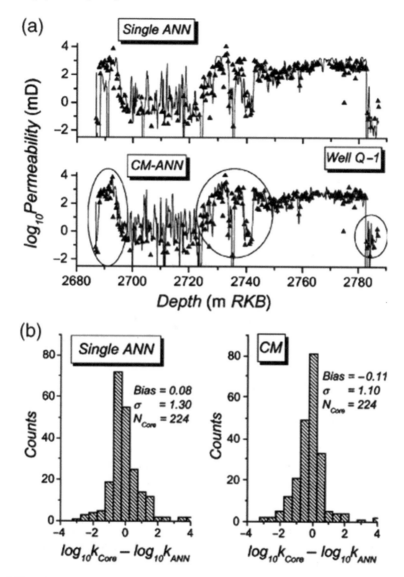

**Fig. 5.12  a** A comparison of permeability predicted by a single ANN and CM ANN. The circles show the sections of improvement. **b** Error distributions from both ANN models (after Bhatt and Helle 2002) (Permission granted from the EAGE Publications)

Recently, Zhong et al. (2019) used CNN to predict permeability in a sandstone reservoir for a carbon sequestration project in West Virginia. The primary focus of their analysis was the Gordon Stray intervals in the Jacksonburg-Stringtown oilfield. They used five variables, such as gamma-ray, bulk density, the slopes of gamma-ray and bulk density curves, and shale volume to generate a stacked feature image, which

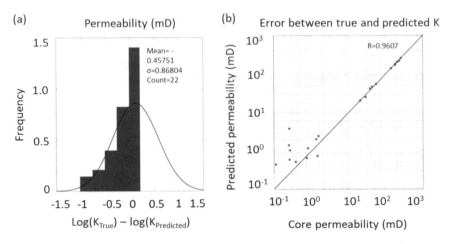

**Fig. 5.13** CNN-predicted permeability in a formation (Zhong et al. 2019). Also see Fig. 3.2 in Chapter 3 from the same study (Permission granted from SEG)

was used to train the fully connected CNN model for permeability prediction. The CNN model consisted of two convolutional layers and two fully connected layers, without any pooling layer. The model showed $R^2$ between CNN-based permeability and ~0.92, with MAE and AAE being 69.7293 and 12.1739 for the test data, which is significantly good (Fig. 5.13). They also showed that CNN performed better than the genetic algorithm-back propagation neural networks.

Fluid saturation is another important petrophysical parameter. Apart from direct onsite fluid sampling and lab-based measurements, we use several conventional log-based empirical models to estimate water saturation (e.g., Archie, Simandoux, Waxmann-Smits, etc.). Apart from specific petrophysical parameters, these models use porosity and resistivity logs for saturation estimates. Each of these empirical models applies to certain reservoir conditions, and they have their own assumptions. Archie's equation applies to porous sandstone reservoirs, whereas the Simandoux method is more applicable to low-resistivity shaly-sand reservoirs often found in deltaic settings. However, it requires the knowledge of cementation factor, saturation exponent, shale resistivity, etc. We can either measure these parameters in the lab or assume certain values based on analogs, none of which are always possible. Carbonate reservoirs pose another set of problems due to the fabric, pore size and type. We can use data-driven techniques in water saturation estimates.

Bhatt (2002) implemented MLP neural network on synthetic and real well logs (resistivity, density, neutron porosity, and sonic) for water saturation estimates in the North Sea. Because of multi-phase fluid in the subsurface, Bhatt also applied a committee machine for each fluid type (oil, gas, and water), with each network consisting of several individually trained neural networks connected in parallel. This model architecture resulted in an overall error reduction by order of magnitude. Khan et al. (2018) applied ANN and ANFIS to predict water saturation using conventional logging suites. The ANN model (with one hidden layer and 20 neurons with an epoch of about 6,000) showed $R^2$ of about 0.92, with an MSE of 0.07, whereas the ANFIS model showed slightly better performance, with an $R^2$ of about 0.96, with a similar MSE. Oruganti et al. (2019) used a tree-based algorithm (XGBoost) to predict gas saturation in a tight gas field in North America.

### 5.2.7.2  Total Organic Carbon Prediction

Total organic carbon (TOC) is one of the most important properties to evaluate the source rock potential. We measure TOC in the lab using different pyrolysis techniques, such as Rock–eval and Hawk pyrolysis. Because of the paucity of core samples, geoscientists have proposed different empirical equations to estimate TOC from conventional well logs over the years (Schmoker and Hester 1983; Passey et al. 1990; Bowman 2010). Although these methods have been widely used across the unconventional plays, they have several limitations due to certain assumptions. Schmoker's method (1983) that uses a density log for TOC estimates assumes any change in the bulk density is due to the presence or absence of low-density kerogen. This equation does not consider the effect of thermal maturity on kerogen properties. Passey's method (1990) uses sonic and resistivity logs for TOC estimates. However, these methods assume similar rock composition, texture, and compaction of the shale formation, which is not true. In addition, the use of the level of maturity (LOM) is another weakness of this technique since it is an uncommon measure (Wang et al. 2016). In addition, the sonic log is not always applicable to correct TOC estimates. Zhu et al. (2019) replaced the sonic log with a gamma-ray log to estimate TOC using Passey's technique, which provided a better match with core data in the Marcellus Shale. This implies that the original Passey's technique is not applicable to all the plays worldwide.

Tan et al. (2015), Mahmoud et al. (2017), and Zhu et al. (2019) used ML for TOC estimates using conventional wireline logging suites. Tan et al. (2015) performed a systematic study to estimate TOC in the Jiumenchong Shale in China using various Support Vector Regressor algorithms (such as Epsilon-SVR, Nu-SVR, and SMO-SVR). They started with a combination of gamma-ray, resistivity, density, photoelectric, neutron porosity, sonic, uranium, potassium, and thorium content to estimate TOC. Their study showed that SVR could be used successfully to estimate TOC with an accuracy of about 83% and MAE of 0.78. They found the RBF function working as the best kernel with their dataset. They also showed model performance with different combinations of well logs and found that the drop of the sonic log results in a significant drop in model performance. SVR-based TOC performed better than the Passey's method. Figures 5.14 and 5.15 show the results from the study. Recently, Zhu et al. (2019) applied deep learning to predict TOC in the Longmaxi and Wufeng formations in the Sichuan Basin of China using conventional logging suites. Because well log interpretation is a problem that involves small sample size, and traditional deep learning with strong feature extraction ability cannot be directly used in such cases; Zhu et al. (2019) used a combination of unsupervised learning and semi-supervised learning in an integrated deep learning model. The model uses a small

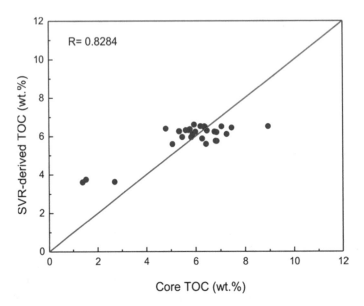

**Fig. 5.14** A comparison of the optimal SVR-based TOC prediction results with core data (Tan et al. 2015) (Reprinted from Journal of Natural Gas Science and Engineering, 26, M Tan, X Song, X Yang, and Q Wu, Support-vector-regression machine technology for total organic carbon content prediction from wireline logs in organic shale: A comparative study, 792–802, Copyright (2015), with permission from Elsevier)

number of labeled samples to build a complex neural network model with more hidden layers for TOC prediction. The model showed better performance (with an MSE of 0.85) than many classic ML models, such as generalized neural network, back-propagation neural network, and random forest, etc.

### 5.2.7.3  Geomechanical Property Prediction

Understanding geomechanical properties is critical to the successful exploitation of tight reservoirs (including enhanced geothermal systems), which need hydraulic stimulation. Brittle materials fracture, whereas ductile materials do not fracture upon applied stress. This depends on the mineral composition, organic matter, and porosity in the rock. Geomechanics also plays a critical role in conventional reservoirs, such as wellbore stability, etc. In general, the important geomechanical properties include minimum horizontal stress, uniaxial compressive stress, Young's modulus, and Poisson's ratio, etc. Ideally, we can measure these properties in a lab under subsurface conditions. However, we do not measure these properties using core samples because of the time, expense (hundreds to thousands of dollars for full core characterization under static and dynamic conditions), and the issue of non-representativeness of the core plug to the formation heterogeneity, which would require a large number of

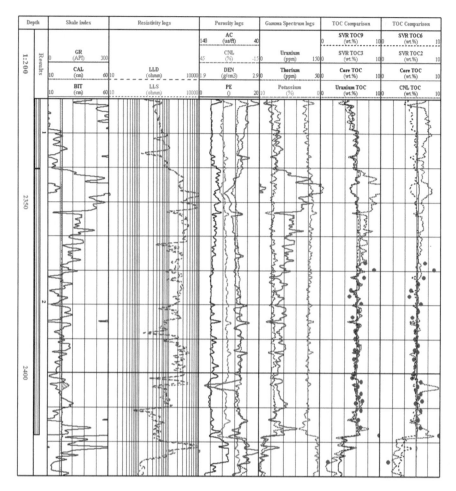

**Fig. 5.15** SVR-based TOC prediction with different wireline logs as model inputs. The SVR-based results with nine- and six-log inputs (TOC9 and TOC6) are consistent with the core-based TOC data, compared to the SVR model with two- or three-log inputs (TOC2 and TOC3) (after Tan et al. 2015) (Reprinted from Journal of Natural Gas Science and Engineering, 26, M Tan, X Song, X Yang, and Q Wu, Support-vector-regression machine technology for total organic carbon content prediction from wireline logs in organic shale: A comparative study, 792–802, Copyright (2015), with permission from Elsevier)

core plugs. The common practice is deriving these properties using multi-component sonic logs (such as dipole sonic). Dipole sonic logs provide both P-wave and S-wave velocity information. However, operators do not always collect these advanced logs due to the cost and time.

Mohaghegh (2017) used ANN to predict geomechanical properties using various sets of conventional well logs drilled through the Marcellus Shale in the Appalachian Basin of the United States. They generated synthetic geomechanical logs for 50 wells out of 80 wells in the study area, which did not have such logs before (Fig. 5.16). Based on the availability of different sets of conventional logs (such as gamma-ray, sonic porosity, and bulk density) in different wells, they generated different groups of wells for the implementation of ANN. In addition to these limited logs, they also added location, depth, and general facies information to constrain the results. This makes sense because the sonic log responses and derived geomechanical properties will vary depending on the Marcellus Shale's depth and location in the basin and its facies variation. After performing the blind tests and evaluating the model results for each well, Mohaghegh used those logs to generate maps to see the lateral variation of geomechanical properties in the study area (Fig. 5.17). We can use ML for this kind of problem for efficient resource recovery, especially in mature basins, where we have various logging data collected over the decades.

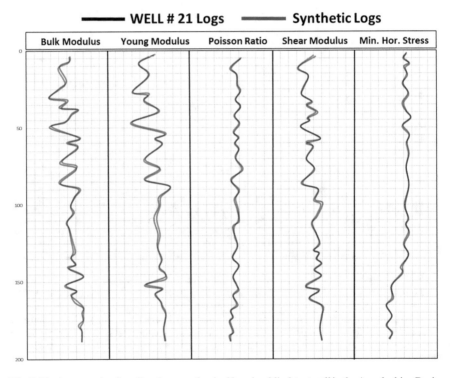

**Fig. 5.16** An example of predicted geomechanical logs in a blind-test well in the Appalachian Basin, United States (after Mohaghegh 2017) (Reprinted/adapted by permission from Springer Nature Customer Service Centre GmbH: Springer Nature, Shale Analytics, Synthetic Geomechanical Logs by SD Mohaghegh © 2017)

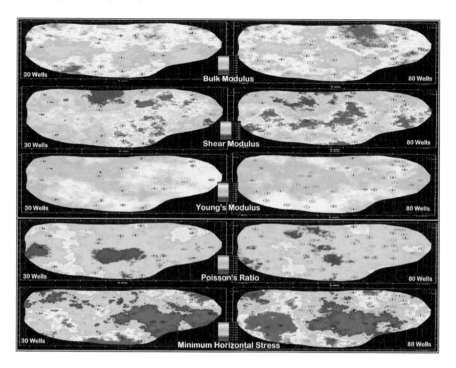

**Fig. 5.17** Distribution of geomechanical properties in the Marcellus Shale before and after ML-modeling. The figures on the left side represent the maps generated using data from 30 wells, whereas the figures on the right side represent the maps generated using synthetic geomechanical logs from 80 wells (after Mohaghegh 2017) (Reprinted/adapted by permission from Springer Nature Customer Service Centre GmbH: Springer Nature, Shale Analytics, Synthetic Geomechanical Logs by S.D. Mohaghegh © 2017)

### 5.2.7.4   Missing Log Prediction and Log Reconstruction

We can use ML to predict missing logs and process existing logs, such as removing noise. Here is an example from a well in the Umiat area on the North Slope, Alaska. The problem was to determine accurate fluid saturation from well log data; however, we can observe several spikes on most of the log curves in the well (Fig. 5.18). The density log is particularly bad (not due to borehole washout in this case), which cannot be used for any meaningful petrophysical analysis. The blue curve shows the original curve (available data), and the black curve shows the despiked curves after filtering.

Basic filtering was applied to the original well logs in the well, but the results did not improve much. ML was implemented to predict a good-quality density log in the well using good-quality gamma-ray, neutron porosity, photoelectric, and sonic logs. A single hidden layer neural network was used for this purpose. For this study, well logs from two nearby wells (within a kilometer) were used to build a ML model that can learn the relation between the good-quality density log (i.e., output) and all

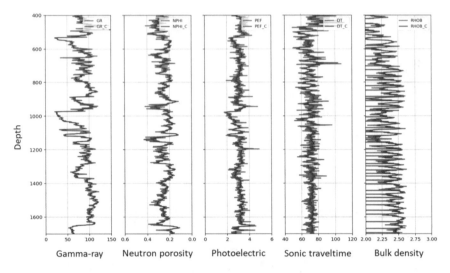

**Fig. 5.18** Conventional wireline logs from a well in northern Alaska. Red curves represent the original log data, and the blue ones indicate the despiked data after filtering. Bulk density curve is particularly bad for any meaningful petrophysical analysis

other well logs (i.e., input) and predict the desired output (i.e., density log) in the target well. Figure 5.19 shows the reconstructed density log for the target well. The

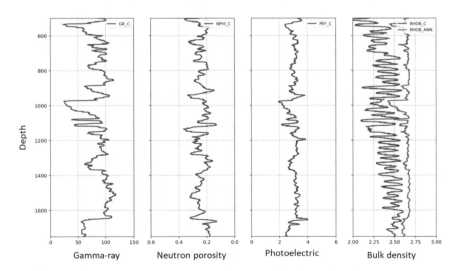

**Fig. 5.19** A well log display showing the despiked logs such as gamma (first track) and neutron porosity (second track), photoelectric (third track), original density and predicted density (both in fourth track) logs for the target well. ML-predicted density log is more meaningful than the original density log

refined density log was used in computing the average porosity for hydrocarbon/water saturation estimations.

## 5.3 Seismic Data Analysis

We can apply ML algorithms to seismic data for structural and stratigraphic interpretations, and quantitative analyses, such as prediction of porosity, TOC, and geomechanical properties. We use both seismic attributes and original seismic amplitude data for building ML models.

### 5.3.1 General Rationale behind the use of Seismic Attributes in ML Applications

Over the years, geophysicists have been developing and using seismic attributes to visualize the subsurface and interpret geology. Seismic attributes are mathematical and statistical quantities derived from the original seismic amplitude. There are attributes and attributes. Some are useful, some complementary, some redundant, and some useless. Marfurt (2018) compiled a list of useful seismic attributes, which we can put into a taxonomic order, such as complex trace, geometric, texture, impedance, and anisotropy attributes. Chen and Sidney (1997), Liner (2004), and Brown (2011) also proposed other classifications of seismic attributes. We can use seismic attributes as input to ML models to classify and predict features of interest. Here we discuss a few specific attributes which have been used in various ML problems.

Coherence: Coherence or the inverse of variance attribute measures the similarity of the waveform. If waveforms show the difference in characteristics, the coherence value will be low, otherwise high. This helps illuminate the boundaries of different geologic features, such as channels, salt, faults, etc.

Curvature: Curvature is the second-order derivative. It measures the degree of curvedness of seismic reflectors. There are different types of curvature algorithms (most-positive and most-negative), which are useful to classify features such as fold, faults, and channel boundaries, etc.

Flexure: Flexure or aberrancy is the third-order derivative. This attribute helps in measuring the lateral change or gradient of curvature along a surface. It is a very helpful attribute to visualize the interactions among various fault segments, especially in an environment affected by multiple fault systems (tectonic and polygonal).

GLCM: The gray-level co-occurrence matrix (GLCM) attributes characterize the local distribution of seismic texture in various statistical ways. There are at least seven different GLCM attributes, including contrast, homogeneity, energy, entropy, similarity, and semblance, etc. A suite of GLCM attributes is useful in seismic facies analysis.

Acoustic Impedance: Relative acoustic impedance and absolute acoustic impedance derived through the seismic inversion process are good facies classifiers and rock property predictors.

$V_P/V_S$ ratio: The ratio of P-wave and S-wave velocity is fundamental; it is related to facies and reservoir property variations. We can derive this attribute by using dipole sonic logs, empirical equations, and inverting seismic data.

Lame's parameters: Lame's parameters, such as lambda-rho (incompressibility) and mu-rho (rigidity), can be used to characterize the elastic properties of the formations. We can feed these attributes to the ML model to characterize the geomechanical properties (and 'fracability') of the tight formations to explore resources, including geothermal energy.

While using seismic attributes in ML applications, we should keep in mind a few important factors, which are as follows.

1. The seismic expression of many of these attributes is subtle that we cannot detect the differences with our eyes; however, when used as input features to the ML model, the machine can 'see' the local differences.
2. It might be better to avoid the attributes with circular data distribution, such as azimuth and phase, because they may make the predicted features more ambiguous.
3. The local difference in attribute expressions is far more important than the regional-scale differences while drilling a well.
4. It is essential to understand the meaning of the individual seismic attributes and their relations to deciphering the geologic heterogeneities; otherwise, the ML effort using them would be futile and perhaps drive the business to failure.

### 5.3.2 Machine Learning for Seismic Facies Clustering and Classification

Similar to well logs, geoscientists have used 2D/3D seismic data for facies classification, including channels, clinoforms, and mass transport complex. In the case of seismic, we use three approaches. The first one is the unsupervised facies classification or clustering based on an ensemble of seismic attributes (Roy et al. 2013, 2014). In such cases, clustering is controlled by selecting input attributes and the number of desired clusters. Roy et al. (2013) used self-organizing map (SOM) and generative topographic map (GTM) techniques to classify facies in a Mississippian chert reservoir in the United States. Initially, they started the work with an ensemble of seismic attributes, such as GLCM entropy, GLCM heterogeneity, spectral bandwidth, coherence, and P-wave impedance, and classified up to 256 clusters, which they reduced down to three geologically meaningful facies derived from well log and core data. The three geologic facies are tight limestone, fractured and layered chert, and high porosity tripolitic chert, the last one being the sweetspot for resource production. Roy et al. (2014) applied the same technique to carbonate wash in the Veracruz Basin of Mexico. Using four seismic attributes (such as P-impedance, lambda-rho, mu-rho, and $V_P/V_S$ ratio in an unsupervised manner, they classified at least four different facies, including carbonate conglomerate wash (with good reservoir potential), harder limestone conglomerate, clay-rich, and tight carbonate facies (Fig. 5.20). Such work is useful to derisk the prospects.

The second approach is on traditional ML-based supervised seismic facies classification. Bhattacharya et al. (2020) used probabilistic neural network (PNN) to classify submarine slide blocks on the North Slope, Alaska. These deposits are highly irregular in nature and associated with imaging artifacts (such as diffraction tails, etc.). They used coherent energy, similarity, and seismic amplitude to classify and predict the mass-transport deposits. Figures 5.21 and 5.22 show the results.

Deep learning is also being used in seismic facies classification now (Alfarraj and AlRegib 2018; Dramsch and Lüthje 2018; Zhao 2018; Alaudah et al. 2019; Di et al. 2019a). Deep learning algorithms do not explicitly use the known seismic attributes an input; rather, it generates a plethora of features from the given seismic image for facies classification. For this reason, deep learning holds great potential for seismic facies classification as it does not require a set of already-defined attributes as input to classify the facies; however, it is also important to physically interpret such features.

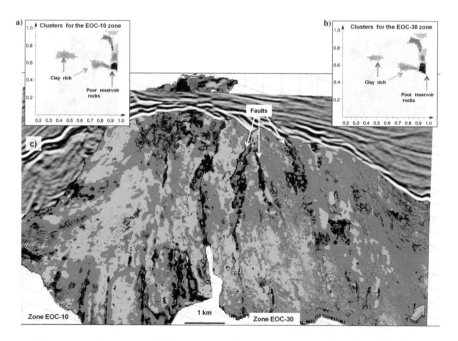

**Fig. 5.20** Seismic facies from GTM clustering within reservoir units (EOC-10 and EOC-30) in a formation in the Veracruz Basin, southern Mexico (after Roy et al. 2014). Seven different polygons with different colors indicate different rock types for reservoir units **a** EOC-10 and **b** EOC-30. **c** The horizon probe generated for the EOC-10 and EOC-30 reservoir units after unsupervised GTM-assisted clustering (Permission granted from SEG)

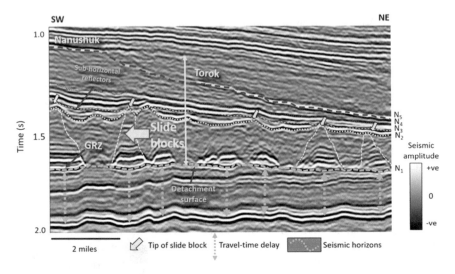

**Fig. 5.21**  A seismic section showing the interpreted submarine slide blocks in a shelf-marine setting in northern Alaska (after Bhattacharya et al. 2020) (Permission granted from SEG)

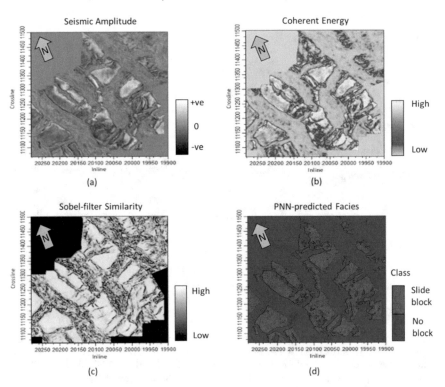

**Fig. 5.22**  Plan views of seismic attributes and attribute-derived slide block distribution using PNN algorithm (modified after Bhattacharya et al. 2020) (Permission granted from SEG)

Zhao (2018) used both fully connected CNN and encoder-decoder CNN to classify facies in the North Sea using F3 seismic dataset. For the fully connected CNN, Zhao (2018) extracted patches of seismic amplitudes around several seed points used in training the model. For the encoder-decoder approach, Zhao manually interpreted a few seismic sections entirely before using them for training. Zhao (2018) used 90% of the data for training and the remaining 10% for testing. The encoder-decoder approach showed better results than the patch-based model (Fig. 5.23). This difference in the model performance may have to do with the approach themselves.

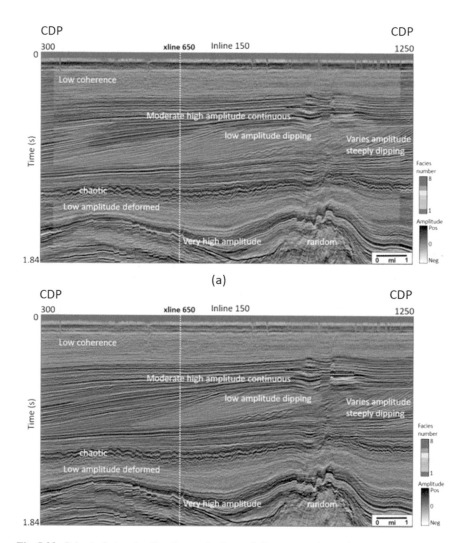

**Fig. 5.23** Seismic facies classification results from a fully connected convolutional neural network **a** and an encoder-decoder convolutional neural network **b** along an inline in the F3 North Sea dataset (Zhao 2018) (Permission granted from SEG)

In the first approach, it is conceivable that geoscientists will pick the minimum number of seed points in the seismic data with the highest quality with the reduced chances of misinterpretations or mislabeling of geologic features, whereas, in the latter approach, we would expect the labeling done using geologic insights (regardless of ambiguities). This process increases the number of training samples that are true representatives of the actual data and helps make the model more generalized than memorized.

Salt body interpretation is another branch of seismic facies classification where ML has been applied for quite some time. Salt body interpretation is important to hydrocarbon exploration, carbon sequestration, and hydrogen storage, as salt bodies provide good seals for the fluids. Compared to regular sedimentary facies, salt bodies are highly irregular in nature and can change its shape both vertically and laterally. Besides, there is often noise (such as migration artifacts) associated with imaging such features. Di and AlRegib (2020) used 2D/3D CNN and compared its performance over MLP-ANN to detect the salt body in a supervised manner in the SEAM seismic dataset (Fig. 5.24). Their CNN model consisted of two convolutional layers, a pooling layer, and a fully connected layer. CNN model showed better results (97% accuracy with a false-positive rate of 0.01) than the MLP (2% overall accuracy with a false-positive rate of 0.08). The pooling layer is useful to reduce overfitting. In comparison to MLP, the CNN model showed faster convergence and more generalization. Sen et al. (2019) used encoder-decoder CNN (U-net) to predict the salt bodies in the Gulf of Mexico.

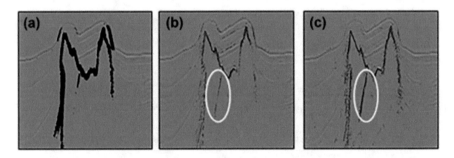

**Fig. 5.24** The comparison of the delineating saltbody boundaries (black), using the sample-level multilayer perceptron (MLP) algorithm from several seismic attributes, including **a** the 9 manually selected seismic attributes, **b** the 8 first-layer convolutional neural network (CNN) attributes, and **c** the 16s-layer CNN attributes (Di and AlRegib 2020). CNN results are better than the MLP algorithm (note the area denoted by ovals) (Permission granted from the EAGE Publications)

### 5.3.3 Fault Classification

Geophysicists have conducted fault classifications in seismic data using ML algorithms (Wu et al. 2018, 2019; Di et al. 2018; Di et al. 2019b; Guitton 2018; Bhattacharya and Di 2020). Similar to facies classification, fault classification is a binary problem. It is simpler than the facies classification problem because it involves only two classes unless we are interested in a deeper understanding of a polyphase fault system. Polyphase fault systems involve several episodes of faulting, and these are highly complex. Such fault systems are present in different sedentary basins globally, including the North Slope, Appalachian Basin, and Permian Basin in the United States. Regardless, the ML-assisted fault classification problem deals with an imbalanced dataset (fault versus no-fault). The number of samples not affected by faults is several magnitudes less than those without being affected by the fault, making it an imbalanced dataset. See Chapter 3 for more details on possible strategies for dealing with imbalanced data.

Di et al. (2019b) used both seismic-attribute-assisted traditional ML and CNN algorithms for fault detection in New Zealand. Perhaps, this is one of the first studies that showed the comparative performance of these two distinct types of models and also marked the transition of research in this area from traditional ML to CNN. They used fourteen seismic attributes as input to the SVM and MLP-NN models for fault model building. The fourteen seismic attributes belonged to primarily three categories, including geometric, edge-detection, and texture attributes. The sample-based MLP and SVM test models showed the overall accuracies of 75% and 72%, respectively. By using the concept of the super attribute (based on local patterns rather than the actual sample values), they increased the accuracies of SVM and MLP up to 88% and 80%, respectively. MLP algorithm showed higher true positives than the SVM model, which is critical to model performance evaluation. Figure 5.25 shows the results from SVM and MLP. Their analysis showed that variance, geometric fault probability, and GLCM contrast attributes contributed the most to fault detection. Although this relative ranking of seismic attributes makes physical sense, not all of them are available in the commercial software packages, and attribute expressions depend on the actual data quality. The other challenge is preparing and maintaining large volumes of 3D seismic attributes in different projects over time and applying them to various ML models with different user-specific choices.

Bhattacharya and Di (2020) applied CNN in two large 3D seismic surveys, covering an area ~1,049 sq. km, to elucidate a polyphase complex fault network on the North Slope, Alaska. It is known that this part of the world underwent several episodes of tectonic deformation over the geologic time period, including at least three rifting events. Each of these events imprinted the subsurface with faults along different orientations (WNW–ESE, E–W, and N–S), corresponding to the stress direction (Fig. 5.26). They used a fully connected CNN model with two convolutional layers, followed by two fully connected layers. They added the dropout technique to avoid model overfitting by preventing complex co-adaptations on training data (Hinton et al. 2012). After model building, they tested the predictions on a few seismic

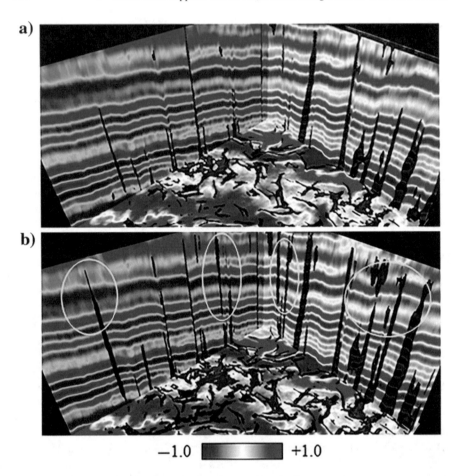

−1.0 ▓▓▓▓▓▓ +1.0

**Fig. 5.25** 3D view of the detected faults in a 3D seismic dataset in New Zealand using the super-attribute-based **a** SVM and **b** MLP classification (Di et al. 2019b). MLP results indicate higher number of faults identified by the MLP algorithm than the SVM algorithm (denoted as the circles) (Permission granted from SEG)

sections and predicted the fault volume throughout both 3D surveys. The overall accuracy was for the test dataset varied between 88.5% and 99.2% for faults only in both surveys. A detailed analysis based on their CNN model also revealed which faults are younger than others and inherited underlying structures (Fig. 5.27). This is indicative of a polyphase fault network. The results were later used in building 3D fault models.

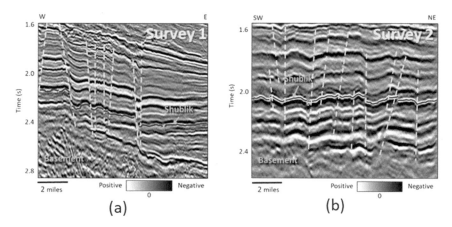

**Fig. 5.26** A seismic section showing the faults from two 3D surveys in northern Alaska (Bhattacharya and Di 2020) (Permission granted from SEG)

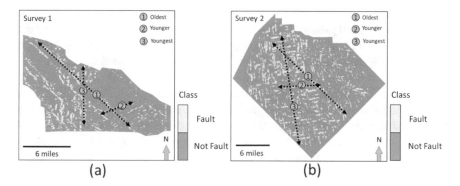

**Fig. 5.27** Plan view of the results from CNN-based faults on the Shublik surface on two 3D seismic surveys in northern Alaska (Bhattacharya and Di 2020). The arrows show the predominant directions of the faults and their cross-cutting nature (Permission granted from SEG)

### 5.3.4 Seismic-Based Rock Property Prediction

There have been several applications of ML for predicting reservoir and geomechanical properties, such as porosity, permeability, total organic carbon, and brittleness index. ML-based solutions are useful for drilling development wells for resource extraction and fluid storage.

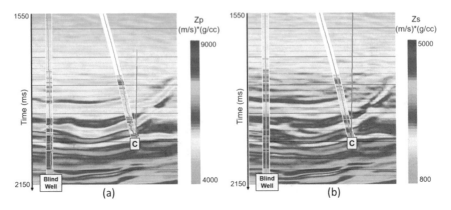

**Fig. 5.28** An example **a** P-wave and **b** S-wave impedance from a deep feed-forward neural network (Dowton et al. 2020) (Permission granted from SEG)

### 5.3.4.1   Acoustic Impedance and Porosity Prediction

Hampson et al. (2001) showed one of the earliest examples of modern-day artificial intelligence for porosity prediction from seismic data. They used MLFN and PNN to predict porosity from multiple seismic attributes. They demonstrated their application in the Blackfoot area of western Canada and the Pegasus field (a Devonian reservoir) of West Texas. In western Canada, they were looking for ways to differentiate between sand-fill and shale-fill sequences in an incised valley-type setting of the Manville group, and in West Texas, they were trying to identify high-porosity reservoir in an anticlinal setting. Their results showed that the combined application of multiple seismic attributes (using multilinear regression) could outperform a single attribute, i.e., acoustic impedance-based porosity prediction. With the increasing number of seismic attributes, the average error of the ML models decreased exponentially. Using ML, we can predict different rock properties (such as porosity, fluid saturation, and clay volume, etc.) in such cases. Hampson et al. (2001) described the methods that could successfully extract several rock properties of interest with higher resolution than conventional seismic data and extend the application of quantitative seismic interpretation techniques to a large extent. This is particularly important, where we have sufficient well control. Dowton et al. (2020) showed the application of deep feed-forward neural network (DNN) to model porosity from seismic data (Figs. 5.28 and 5.29).

### 5.3.4.2   Total Organic Carbon and Brittleness Prediction

Verma et al. (2016) performed an important study predicting TOC and brittleness index (BI) volumes using core, well logs, and 3D seismic data in the Barnett Shale in North America. They hypothesized that TOC and BI could be derived from seismic data, and brittleness is related to mineralogy that could be predicted using well logs

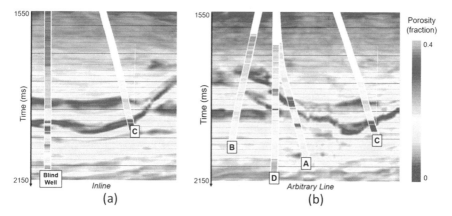

**Fig. 5.29** DNN-based porosity estimates at **a** the inline going through the blind well and **b** the arbitrary line going through the four wells that were initially used to generate the pseudowells (Dowton et al. 2020) (Permission granted from SEG)

and seismic data. First, they predicted TOC using several well logs to match it to core-based TOC measurements. Then, they used log-estimated TOC from 30 wells, 3D seismic-inverted P-impedance, S-impedance, lambda-rho, mu-rho, relative acoustic impedance, total energy, and stratigraphic height to model the TOC and brittleness volumes using multilinear regression and probabilistic neural network (PNN). The PNN-model-based prediction accuracies of TOC and brittleness index were 87% and 68%, which was a little higher than the multilinear regression technique. See Fig. 5.30.

## 5.4   Fiber-Optic-Based Fluid Flow Prediction

Fiber-optic is an advanced non-invasive and dynamic subsurface monitoring tool, which can record temperature and strain around the well (Ghahfarokhi et al. 2018). This technology uses a telecommunication cable either set up on the surface or in the subsurface. Distributed acoustic sensing (DAS) and distributed temperature sensing (DTS) are two fiber-optic tools that measure the strain or strain rate, and temperature in the well, respectively. These tools provide continuous multipoint reservoir temperature and strain monitoring, which can be used in seismic, microseismic data acquisition and fluid flow monitoring, for oil and gas exploration, geothermal resources, and earthquake seismology (Binder and Tura 2020; Stork et al. 2020). DAS datasets are generally available in the SEG-Y format, like seismic data, whereas DTS data are available in ASCII or CSV format. Depending on the recording intervals and the total length of recording, the data size from fiber optic can be truly big data, reaching up to petabytes. Processing, displaying, and identifying important features from such data are hard using currently available resources. Because the acquisition of such data

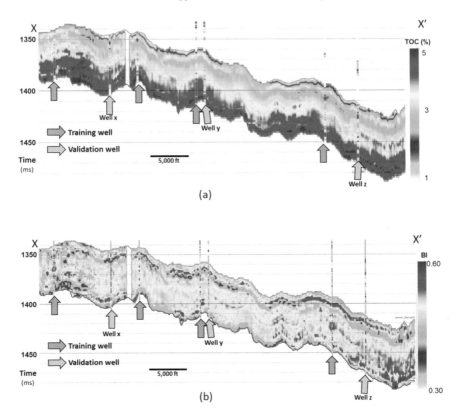

**Fig. 5.30** **a** A vertical slice along the line XX' through ML-based total organic carbon TOC volume. **b** A vertical slice along the line XX' through ML-based BI volume of the Lower Barnett (Verma et al. 2016). Some of the layers have high TOC and high brittleness, which could indicate potential sweet spots (Permission granted from SEG)

in the subsurface is expensive, we can use data analytics to glean meaningful and actionable information from such data. By deploying ML on DTS data from each hydraulically fractured stage in a 28-stage horizontal well in the Marcellus Shale, Ghahfarokhi et al. (2018) could predict gas production. Figure 5.31 shows DTS and fluid production data from a horizontal well in the Marcellus Shale, and Fig. 5.32 shows the comparison of actual versus ML-predicted daily gas production from individual completion stages. Ghahfarokhi et al. (2018) used upscaled DTS data and flowtime from previous completion stages to predict gas production for each consecutive stage. It is a time series problem. Their sensitivity analysis showed that certain stages are more productive than other stages, which are tied to the rock properties, such as Poisson's ratio. Rocks with a low Poisson's ratio are more brittle than rocks with a high Poisson's ratio, which affects the efficacy of hydraulic stimulation and fracturing. Stages that were hydraulically stimulated using common engineering or geometric approach were found to be less productive than geo-engineered stages.

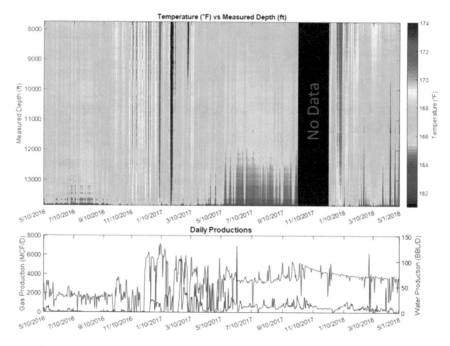

**Fig. 5.31** An example of DTS signal and fluid production (gas and water) data from a well in the Marcellus Shale (Ghahfarokhi et al. 2018) (This figure reprinted from PK Ghahfarokhi, TR Carr, S Bhattacharya, J Elliott, A Shahkarami, and K Martin, 2018, with permission from URTeC, whose permission is required for further use)

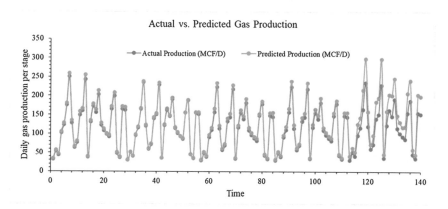

**Fig. 5.32** An example of actual versus predicted daily gas production from individual completion stages from a horizontal well in the Marcellus Shale. Random Forest algorithm was used for the study. Initially, the model works well; however, the model performance deteriorates after some time

It also indicates that the shale formations are laterally heterogeneous, and therefore, hydraulic stimulation operations will have to take these geologic controls into account for efficient production.

## 5.5  Rock Characterization (Core, Outcrop, Petrography, and Geochemistry)

Although the published studies on ML applications in subsurface are more in well logs and seismic data than the actual rocks and core data, such ML-based studies are becoming more and more common. As rocks represent the ground-truth, ML's application has a very strong promise for fundamental geologic characterization and predictive modeling, which can be useful in resources exploration, including hydrocarbons and minerals. This is true, especially in those formations where we have gathered data from cores and cuttings, including geochemistry (major and trace elements, and isotopes), grain density, porosity–permeability, and fluid saturation, etc.

Recent studies showed the application of deep learning to analyze high-resolution images, such as thin sections, CT scan, micro-CT scan, SEM, and fossil identifications (Duarte-Coronado et al. 2019; Pires de Lima et al. 2019, 2020; Wang et al. 2019; Alqahtani et al. 2020). ML can be used to automate the identification of fossils, minerals, organic matter, porosity, and fracture identification. Many geological surveys and museums are involved in digital archiving of their data these days. For example, as biostratigraphy is becoming of lesser emphasis in paleontology, the number of paleontologists is not growing like before (at least in North America), and there is a major change in the workforce is ongoing, a lot of the valuable knowledge and expertise can be lost if not digitally captured and archived for future use.

Pires de Lima et al. (2019) showed an example of core-based lithofacies classification using convolutional neural network. They used a 700-ft (~213 m) slabbed core from the Mississippian limestone and chert reservoirs in the Anadarko Shelf in Oklahoma. They used expert-labeled 17 different lithofacies on the core slabs, including wackestone, grainstone, mudstone, and sandstone, etc. for supervised facies classification (Fig. 5.33). They used different network architectures, such ResNet, Inception, MobileNet, and NasNet. It is known that the first few layers of a deep neural network can learn features that are useful to identify textures or colors, and often lithofacies descriptions are based on these primary characteristics. The results from the CNN networks produced similar results, with 90–95% accuracy. As it is well-known that overall model accuracy is just one of the metrics to evaluate model performance, it is not necessarily useful in several cases, where the granularity of analysis is more important, such as facies. ML may have different levels of performance for each facies. Based on that, Pires de Lima et al. (2019) assigned a particular threshold of probability (0.3) for either accepting or rejecting the results. They selected this threshold of 0.3 so that all images would be classified. One of the major utilities of

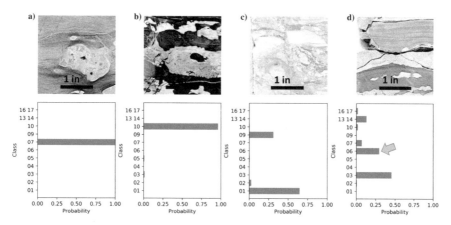

**Fig. 5.33** Examples of the classification performed by the retrained ResNetV2 network (Pires de Lima et al. 2019). **a** Nodular packstone-grainstone (facies 7), **b** bioturbated mudstone-wackestone (facies 10), **c** chert breccia (facies 1) and bioturbated skeletal peloidal packstone-grainstone (facies 9), and **d** bedded skeletal peloidal packstone grainstone (facies 6). CNN failed to accurately classify facies 6 (Permission granted from SEG)

this type of work is that a sedimentologist can describe facies in certain intervals and then use CNN technique to predict facies using the cored images throughout the whole interval and then quality-check the prediction and update the ML model as needed. Pires de Lima et al. (2020) also demonstrated an innovative example of fossil identification using convolutional neural network. Nanjo and Tanaka (2019) used CNN to predict 306 petrographic thin sections to identify carbonate lithofacies in Japan.

Geochemistry is another branch that is ripe for ML applications. Currently, ML is used in geochemistry for dimensionality reduction, rock classification, detection of geochemical anomalies, and mapping (Kuwatani et al. 2014; Chen et al. 2014; Zuo et al. 2019; Duarte et al. 2020). For example, a hand-held XRF instrument yields numerous major and trace element composition. We can use different statistical techniques to reduce the large data-dimensionality and generate relevant features for chemofacies classification. Duarte et al. (2020) used ten features from a large XRF dataset from Oklahoma for unsupervised clustering using hierarchical cluster analysis (HCA), K-means, and DBSCAN methods (Fig. 5.34). They used the method to identify the optimal number of clusters based on the sum of the squared distance. They compared the XRF-derived chemofacies with petrographic thin sections and wireline logs. This helps in inferring the ocean chemistry, depositional and diagenetic environment, and paleo-anoxia. Milad (2019) and Milad et al. (2020) extended the application of ML (e.g., self-organizing map) and XRF data from core to outcrops and they correlated electrofacies, chemofacies, and lithofacies between outcrop and subsurface samples (Fig. 5.35). Often core samples are missing from a certain logged interval where we can deploy ML on the existing XRF data to build a non-linear regression model to generate pseudo-elemental logs, similar to missing petrophysical

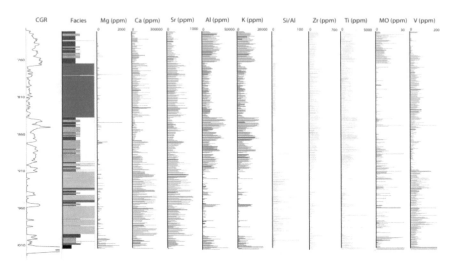

**Fig. 5.34** Vertical profile of elemental composition with gamma ray (CGR) and facies in the Woodford shale, Osage, and Meramec intervals in the Anadarko Basin, Oklahoma (after Duarte-Coronado et al. 2019) (Permission received)

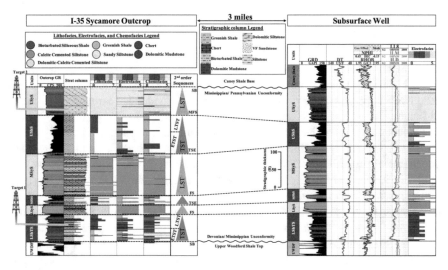

**Fig. 5.35** Outcrop-to-subsurface stratigraphic correlation of the Mississippian Sycamore interval in Oklahoma, United States (after Milad et al. 2020). Drilling rigs on the left side are the proposed horizontal wells within two Mississippian Sycamore sections (Reprinted from Marine and Petroleum Geology, 115, B Milad, R Slatt, and Z Fuge, Lithology, stratigraphy, chemostratigraphy, and depositional environment of the Mississippian Sycamore rock in the SCOOP and STACK area, Oklahoma, USA: Field, lab, and machine learning studies on outcrops and subsurface wells, 1–18, Copyright (2020), with permission from Elsevier)

log reconstruction. We can also use ML to map XRF-based elemental data to XRD-based mineral composition (Alnahwi and Loucks 2019). These problems are based on numerical data, and we can use traditional ML algorithms for most of these problems. Chemistry-informed ML holds a tremendous promise in crystallography and geochemistry, such as simulating rock-fluid-fracture interactions, nucleation, growth, and mineralization. Leal et al. (2017) used ML to simulate the precipitation of calcite and dolomite due to injection of different fluid along a rock core at three different simulation times. Such applications can have impacts on the exploration of unconventional energy resources, geothermal energy, carbon storage, and hydrogen storage projects.

# References

Abegg, FE, Loope DB, Harris PM (2001) Carbonate eolianites: depositional models and diagenesis. In: Abegg FE, Harris PM, Loope DB (eds) Modern and ancient carbonate eolianites: sedimentology, sequence stratigraphy, and diagenesis. SEPM Special Publication 71, pp 17–30. https://doi.org/10.2110/pec.01.71.0017

Al-Anazi AF, Gates ID (2010) Support vector regression for porosity prediction in a heterogeneous reservoir: a comparative study. Comput Geosci 36(12):1494–1503. https://doi.org/10.1016/j.cageo.2010.03.022

Al-Anazi AF, Gates ID (2012) Support vector regression to predict porosity and permeability: effect of sample size. Comput Geosci 39:64–76. https://doi.org/10.1016/j.cageo.2011.06.011

Alaudah Y, Michalowicz P, Alfarraj M, AlRegib G (2019) A machine learning benchmark for facies classification. Interpretation 7(3):SE175–SE187. https://doi.org/10.1190/INT-2018-0249.1

Alfarraj M, AlRegib G (2018) Petrophysical-property estimation from seismic data using recurrent neural networks. SEG Technical Program Expanded Abstracts, 2141–2146. https://doi.org/10.1190/segam2018-2995752.1

Alnahwi A, Loucks RG (2019) Mineralogical composition and total organic carbon quantification using x-ray fluorescence data from the Upper Cretaceous Eagle Ford Group in southern Texas. Am Asso Petrol Geol Bull 103(12):2891–2907. https://doi.org/10.1306/04151918090

Alqahtani N, Alzubaidi F, Armstrong RT, Swietojanski P, Mostaghimi P (2020) Machine learning for predicting properties of porous media from 2D X-ray images. J Petrol Sci Eng 184:106514. https://doi.org/10.1016/j.petrol.2019.106514

Bhatt A (2002) Reservoir properties from well logs using neural networks. PhD dissertation, Norwegian University of Science and Technology

Bhatt A, Helle HB (2002) Committee neural networks for porosity and permeability prediction from well logs. Geophys Prospect 50(6):645–660. https://doi.org/10.1046/j.1365-2478.2002.00346.x

Bhattacharya S, Carr TR, Pal M (2016) Comparison of supervised and unsupervised approaches for mudstone lithofacies classification: case studies from the Bakken and Mahantango-Marcellus Shale, USA. J Nat Gas Sci Eng 33:1119–1133. https://doi.org/10.1016/j.jngse.2016.04.055

Bhattacharya S, Carr T, Wang G (2015) Shale lithofacies classification and modeling: case studies from the Bakken and Marcellus formations, North America. Presented at American association of petroleum geologists annual conference, Denver, May 31–June 3

Bhattacharya S, Di H (2020) The classification and interpretation of the polyphase fault network on the North Slope, Alaska using deep learning. SEG Technical Program Expanded Abstracts, 3847–3851. https://doi.org/10.1190/segam2020-w13-01.1

Bhattacharya S, Mishra S (2018) Applications of machine learning for facies and fracture prediction using Bayesian Network Theory and Random Forest: case studies from the Appalachian basin, USA. J Petrol Sci Eng 170:1005–1017. https://doi.org/10.1016/j.petrol.2018.06.075

Bhattacharya S, Tian M, Rotzien J, Verma S (2020) Application of seismic attributes and machine learning for imaging submarine slide blocks on the North Slope, Alaska. SEG Technical Program Expanded Abstracts, 1096–1100. https://doi.org/10.1190/segam2020-3426887.1

Binder G, Tura A (2020) Convolutional neural networks for automated microseismic detection in downhole distributed acoustic sensing data and comparison to a surface geophone array. Geophys Prospect 68(9):2770–2782

Bowman T (2010) Direct method for determining organic shale potential from porosity and resistivity logs to identify possible resource play. American association of petroleum geologists search and discovery article #110128

Brown AR (2011) Interpretation of three-dimensional seismic data. Society of exploration geophysicists and the American association of petroleum geologists

Chen Q, Sidney S (1997) Seismic attribute technology for reservoir forecasting and monitoring. Lead Edge 16(5):445–448. https://doi.org/10.1190/1.1437657

Chen Y, Lu L, Li X (2014) Application of continuous restricted Boltzmann machine to identify multivariate geochemical anomaly. J Geochem Explor 140:56–63. https://doi.org/10.1016/J.GEXPLO.2014.02.013

Deng T, Xu C, Jobe D, Xu R (2019) A comparative study of three supervised machine-learning algorithms for classifying carbonate vuggy facies in the Kansas Arbuckle Formation. Petrophysics 60(6):838–853. https://doi.org/10.30632/PJV60N6-2019a8

Di H, AlRegib G (2020) A comparison of seismic saltbody interpretation via neural networks at sample and pattern levels. Geophys Prospect 68(2):521–535. https://doi.org/10.1111/1365-2478.12865

Di H, Gao D, AlRegib G (2019a) Developing a seismic texture analysis neural network for machine-aided seismic pattern recognition and classification. Geophys J Int 218(2):1262–1275. https://doi.org/10.1093/gji/ggz226

Di H, Shafiq MA, Wang Z, AlRegib G (2019b) Improving seismic fault detection by super-attribute-based classification. Interpretation 7(3):SE251–SE267. https://doi.org/10.1190/INT-2018-0188.1

Di H, Wang Z, AlRegib G (2018) Seismic fault detection from post-stack amplitude by convolutional neural networks. Conference proceedings, 80th EAGE conference and exhibition, pp 1–5. https://doi.org/10.3997/2214-4609.201800733

Dong S, Zeng L, Lyu W, Xia D, Liu G, Wu Y, Du X (2020) Fracture identification and evaluation using conventional logs in tight sandstones: a case study in the Ordos Basin, China. Energy Geosci 1(3–4):115–123. https://doi.org/10.1016/j.engeos.2020.06.003

Dowton JE, Collet O, Hampson DP, Colwell T (2020) Theory-guided data science-based reservoir prediction of a North Sea oil field. Lead Edge 39(10):742–750. https://doi.org/10.1190/tle39100742.1

Dramsch JS, Lüthje M (2018) Deep-learning seismic facies on state-of-the-art CNN architectures. SEG Technical Program Expanded Abstracts, 2036–2040. https://doi.org/10.1190/segam2018-2996783.1

Duarte D, Lima R, Slatt R, Marfurt K (2020) Comparison of clustering techniques to define chemofacies in mississippian rocks in the STACK Play, Oklahoma. American association of petroleum geologists search and discovery, 42523. https://doi.org/10.1306/42523Duarte2020

Duarte-Coronado D, Tellez-Rodriguez J, Pires de Lima R, Marfurt KJ, Slatt R (2019) Deep convolutional neural networks as an estimator of porosity in thin-section images for unconventional reservoirs. SEG Technical Program Expanded Abstracts, 3181–3184. https://doi.org/10.1190/segam2019-3216898.1

Ghahfarokhi PK, Carr TR, Bhattacharya S, Elliott J, Shahkarami A, Martin K (2018) A fiber-optic assisted multilayer perceptron reservoir production modeling: a machine learning approach in prediction of gas production from the Marcellus shale. Presented at the SPE/AAPG/SEG unconventional resources technology conference, Houston, Texas. URTEC-2902641-MS. https://doi.org/10.15530/URTEC-2018-2902641

Guitton A (2018) 3D convolutional neural networks for fault interpretation. 80th EAGE conference and exhibition. https://www.earthdoc.org/publication/publicationdetails/?publication=92118

Hall B (2016) Facies classification using machine learning. Lead Edge 35(10):906–909. https://doi.org/10.1190/tle35100906.1

Hampson DP, Schuelke JS, Quirein JA (2001) Use of multiattribute transforms to predict log properties from seismic data. Geophysics 66(1):220–236. https://doi.org/10.1190/1.1444899

Handford CR, Francka BJ (1991) Mississippian carbonate-siliciclastic eolianites in southwestern Kansas. In: Mixed Carbonate-Siliciclastic Sequences, Lomando AJ, Harris PM (eds) Society of economic paleontologists and mineralogists. Core Workshop No 15, pp 205–243. https://doi.org/10.2110/cor.91.01.0205

Helle HB, Bhatt A, Ursin B (2001) Porosity and permeability prediction from wireline logs using artificial neural networks: a North Sea case study. Geophys Prospect 49(4):431–444. https://doi.org/10.1046/j.1365-2478.2001.00271.x

Hinton G, Srivastava N, Krizhevsky A, Sutskever I, Salakhutdinov RR (2012) Improving neural networks by preventing co-adaptation of feature detectors. arXiv:1207.0580

Howat E, Mishra S, Schuetter J, Grove B, Haagsma A (2016) Identification of Vuggy Zones in carbonate reservoirs from wireline logs using machine learning techniques. American association of petroleum geologists eastern section 44th annual meeting. https://doi.org/10.13140/RG.2.2.30165.73443

Ja'fari A, Kadkhodaie-Ilkhchi A, Sharghi Y, Ghanavati K (2011) Fracture density estimation from petrophysical log data using the adaptive neuro-fuzzy inference system. J Geophys Eng 9(1):105–114. https://doi.org/10.1088/1742-2132/9/1/013

Khan MR, Tariq Z, Abdulraheem A (2018) Machine learning derived correlation to determine water saturation in complex lithologies. Presented at the SPE Kingdom of Saudi Arabia annual technical symposium and exhibition, Dammam, Saudi Arabia. SPE-192307-MS. https://doi.org/10.2118/192307-MS

Kuwatani T, Nagata K, Okada M, Watanabe T, Ogawa Y, Komai T, Tsuchiya N (2014) Machine-learning techniques for geochemical discrimination of 2011 Tohoku tsunami deposits. Scientific Reports 4:7077. https://doi.org/10.1038/srep07077

Leal AMM, Kulik DA, Saar MO (2017) Ultra-fast reactive transport simulations when chemical reactions meet machine learning: chemical equilibrium. arXiv:1708.04825

Liner C (2004) Elements of 3D seismology. Investigations in geophysics No 19. Society of exploration geophysicists

Mahmoud AAA, Elkatatny S, Mahmoud M, Abouelresh M, Abdulraheem A, Ali A (2017) Determination of the total organic carbon (TOC) based on conventional well logs using artificial neural network. Int J Coal Geol 179:72–80. https://doi.org/10.1016/j.coal.2017.05.012

Marfurt KJ (2018) Seismic attributes as the framework for data integration throughout the oilfield life cycle. Distinguished instructor short course, Society of Exploration Geophysicists. https://doi.org/10.1190/1.9781560803522

Milad B (2019) Integrated reservoir characterization and geological upscaling for reservoir flow simulations of the Sycamore/Meramec and Hunton plays in Oklahoma. PhD dissertation, University of Oklahoma

Milad B, Slatt R, Fuge Z (2020) Lithology, stratigraphy, chemostratigraphy, and depositional environment of the Mississippian Sycamore rock in the SCOOP and STACK area, Oklahoma, USA: Field, lab, and machine learning studies on outcrops and subsurface wells, Marine and Petroleum Geology, 115. https://doi.org/10.1016/j.marpetgeo.2020.104278

Misra S, Li H, He J (2019) Machine learning for subsurface characterization. Gulf Publishing

Mohaghegh SD (2017) Shale analytics: data-driven analytics in unconventional resources. Springer International Publishing. https://doi.org/10.1007/978-3-319-48753-3

Mohaghegh SD, Ameri S (1995) Artificial neural network as a valuable tool for petroleum engineers. SPE 29220, Society of Petroleum Engineers

Nanjo T, Tanaka S (2019) Carbonate lithology identification with machine learning. Presented at the Abu Dhabi international petroleum exhibition & conference, Abu Dhabi. UAE SPE-197255-MS. https://doi.org/10.2118/197255-MS

Oruganti YD, Yuan P, Inanc F, Kadioglu Y, Chace D (2019) Role of machine learning in building models for gas saturation prediction, SPWLA 60th annual logging symposium

Passey QR, Creaney S, Kulla JB, Moretti FJ, Stroud JD (1990) A practical model for organic richness from porosity and resistivity logs. Am Asso Petrol Geol Bull 74:1777–1794

Pires de Lima R, Suriamin F, Marfurt KJ, Pranter MJ (2019) Convolutional neural networks as aid in core lithofacies classification. Interpretation 7(3):SF27–SF40. https://doi.org/10.1190/INT-2018-0245.1

Pires de Lima R, Welch KF, Barrick JE, Marfurt KJ, Burkhalter R, Cassel M, Soreghan GS (2020) Convolutional neural networks as an aid to biostratigraphy and micropaleontology: a test on late Paleozoic microfossils. Palaios 35(9):391–402. https://doi.org/10.2110/palo.2019.102

Qi L, Carr TR (2006) Neural network prediction of carbonate lithofacies from well logs, Big Bow and Sand Arroyo Creek fields, Southwest Kansas. Comput & Geosci 32(7):947–964. https://doi.org/10.1016/j.cageo.2005.10.020

Rafik B, Kamel B (2017) Prediction of permeability and porosity from well log data using the nonparametric regression with multivariate analysis and neural network, Hassi R'Mel Field, Algeria. Egypt J Pet 26(3):763–778. https://doi.org/10.1016/j.ejpe.2016.10.013

Rcnguang Z, Xiong Y, Wang J, Carranza EJM (2019) Deep learning and its application in geochemical mapping. Earth Sci Rev 192:1–14. https://doi.org/10.1016/j.earscirev.2019.02.023

Rogers SJ, Chen HC, Kopaska-Merkel DC, Fang JH (1995) Predicting permeability from porosity using artificial neural networks 1. Am Asso Petrol Geol Bull 79(12):1786–1797. https://doi.org/10.1306/7834DEFE-1721-11D7-8645000102C1865D

Roy A, Dowdell BL, Marfurt KJ (2013) Characterizing a Mississippian tripolitic chert reservoir using 3D unsupervised and supervised multiattribute seismic facies analysis: an example from Osage County, Oklahoma. Interpretation 1(2):SB109–SB124. https://doi.org/10.1190/INT-2013-0023.1

Roy A, Romero-Peláez AS, Kwiatkowski TJ, Marfurt KJ (2014) Generative topographic mapping for seismic facies estimation of a carbonate wash, Veracruz Basin, southern Mexico. Interpretation 2(1):SA31–SA47. https://doi.org/10.1190/INT-2013-0077.1

Schmoker JW, Hester TC (1983) Organic carbon in Bakken formation, United States portion of Williston Basin. Am Asso Petrol Geol Bull 67:2165–2174

Sen D, Ong C, Kainkaryam S, Sharma A (2020) Automatic detection of anomalous density measurements due to wellbore cave-in. Petrophysics 61(5):434–449. https://doi.org/10.30632/PJV61N5-2020a3

Sen S, Kainkaryam S, Ong C, Sharma A (2019) Regularization strategies for deep-learning-based salt model building. Interpretation 7(4):T911–T922. https://doi.org/10.1190/INT-2018-0229.1

Shazly T, Tarabees EA (2013) Using of Dual Laterolog to detect fracture parameters for Nubia sandstone formation in Rudeis-Sidri area, Gulf of Suez, Egypt. Egypt J Pet 22(2):313–319. https://doi.org/10.1016/j.ejpe.2013.08.001

Stork AL, Baird AF, Horne SA, Naldrett G, Lapins S, Kendall JM, WookeyJ, Verdon JP, Clarke A, Williams A (2020) Application of machine learning to microseismic event detection in distributed acoustic sensing data. Geophysics 85(5):KS149–KS160. https://doi.org/10.1190/geo2019-0774.1

Tan M, Song X, Yang X, Wu Q (2015) Support-vector-regression machine technology for total organic carbon content prediction from wireline logs in organic shale: a comparative study. J Nat Gas Sci Eng 26:792–802. https://doi.org/10.1016/j.jngse.2015.07.008

Tokhmchi B, Memarian H, Rezaee MR (2010) Estimation of the fracture density in fractured zones using petrophysical logs. J Petrol Sci Eng 72(1–2):206–213. https://doi.org/10.1016/j.petrol.2010.03.018

Trainor-Guitton W, Jreij S, Guitton A, Simmons J (2018) Fault classification from 3D imaging of vertical DAS profile. SEG Technical Program Expanded Abstracts, 4664–4668. https://doi.org/10.1190/segam2018-2989447.1

Vasvári V (2011) On the applicability of dual Laterolog for the determination of fracture parameters in hard rock aquifers. Austrian J Earth Sci 104(2):80–89

Verma S, Zhao T, Marfurt KJ, Devegowda D (2016) Estimation of total organic carbon and brittleness volume. Interpretation 4(3):T373–T385. https://doi.org/10.1190/INT-2015-0166.1

Wang G (2012) Black shale Lithofacies prediction and distribution pattern analysis of middle Devonian Marcellus shale in the Appalachian basin. Northeastern U.S.A. PhD thesis, West Virginia University

Wang P, Chen Z, Pang X, Hu K, Sun M, Chen X (2016) Revised models for determining TOC in shale play: example from Devonian Duvernay Shale, Western Canada Sedimentary Basin. Mar Pet Geol 70:304–319. https://doi.org/10.1016/j.marpetgeo.2015.11.023

Wang Y, Teng Q, He X, Feng J, Zhang T (2019) CT-image of rock samples super resolution using 3D convolutional neural network. Comput Geosci 133:104314. https://doi.org/10.1016/j.cageo.2019.104314

Wood DA (2018) A transparent open-box learning network provides insight to complex systems and a performance benchmark for more-opaque machine-learning algorithms. Adv Geo-Energy Res 2(2):148–162. https://doi.org/10.26804/ager.2018.02.04

Wood DA (2019) Lithofacies and stratigraphy prediction methodology exploiting an optimized nearest-neighbour algorithm to mine well-log data. Mar Pet Geol 110:347–367. https://doi.org/10.1016/j.marpetgeo.2019.07.026

Wu X, Liang L, Shi Y, Fomel S (2019) FaultSeg3D: Using synthetic datasets to train an end-to-end convolutional neural network for 3D seismic fault segmentation. Geophysics 84(3):IM35–IM45. https://doi.org/10.1190/geo2018-0646.1

Wu X, Shi Y, Fomel S, Liang L (2018) Convolutional neural networks for fault interpretation in seismic images. SEG Technical Program Expanded Abstracts, 1946–1950. https://doi.org/10.1190/segam2018-2995341.1

Zazoun RS (2013) Fracture density estimation from core and conventional well logs data using artificial neural networks: the Cambro-Ordovician reservoir of Mesdar oil field, Algeria. J Afr Earth Sc 83:55–73. https://doi.org/10.1016/j.jafrearsci.2013.03.003

Zhao T (2018) Seismic facies classification using different deep convolutional neural networks. SEG Technical Program Expanded Abstracts, 2046–2050. https://doi.org/10.1190/segam2018-2997085.1

Zhao, T., Jayaram, V., Roy, A., Marfurt, K. J. (2015) A comparison of classification techniques for seismic facies recognition. Interpretation 3(4):SAE29–SAE58. https://doi.org/10.1190/INT-2015-0044.1

Zhong Z, Carr TR, Wu X, Wang G (2019) Application of a convolutional neural network in permeability prediction: a case study in the Jacksonburg-Stringtown oil field, West Virginia, USA. Geophysics 84(6):B363–B373. https://doi.org/10.1190/geo2018-0588.1

Zhu L, Zhang C, Zhang C, Zhang Z, Nie X, Zhou X, Liu W, Wang X (2019) Forming a new small sample deep-learning model to predict total organic carbon content by combining unsupervised learning with semisupervised learning. Appl Soft Comput 83:105596. https://doi.org/10.1016/j.asoc.2019.105596

Zuo R, Xiong Y, Wang J, Carranza EJM (2019) Deep learning and its application in geochemical mapping. Earth Sci Rev 192:1–14

# Chapter 6
# The Road Ahead

**Abstract** In the last chapter, I discuss the future of data analytics (DA) and machine learning (ML) in geosciences research, instruction, community, and business, as a whole. It sets an agenda of ML-focused studies that need to be conducted to solve critical problems in geosciences. This endeavor will not only help understand the fundamental geologic processes and better analyze rocks but also assist the businesses to make better decisions and grow as needed.

**Keywords** Future of AI · Multi-tasking ML · Physics-informed ML · Rare event analytics · Geosciences education · Energy transition · Perils of ML

It has been about 80 years since the birth of the Turing machine. Since then, artificial intelligence (AI) has flourished, sparked new interests and controversies, gone through at least two winters, and reemerged with new capabilities and applications. The first two waves of AI (1950's and 1980's) were centered around developing new algorithms, and AI was mostly concentrated in the statistics, mathematics, and biology communities. The ongoing third wave of AI is different than the previous two in several ways. It has transcended across all disciplines and fostered new collaborations across businesses. It has facilitated the development of new computational technologies, affordable online courses in AI and general-purpose coding, cloud-based solutions, and the building and release of large datasets to the public. In a way, many of these changes have happened in a bottom-up approach, with customers feeding the AI frenzy, not just the developers of ML algorithms, as in the previous two AI waves. Of course, there are some top-down emphasis on analytics coming from the management side to increase business efficiency. These are some of the fundamental differences wherein lies the direction DA and AI will take in the next several years. We will use data analytics across industry, in research labs, and in higher education in different ways to solve organization-specific problems and provide better solutions to customers.

As we advance, we will see ML becoming more capable of solving complex, dynamic and multitasking problems, more and more businesses using it, new visualization tools, and availability of reproducible research codes and datasets. More and

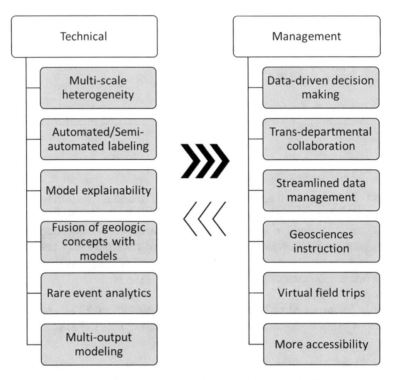

**Fig. 6.1** Major areas in geoscience (technical and managerial) in which ML and DA will have a long-term impact. Many of these improvements will be highly iterative and collaborative

more geoscientists will adopt ML. ML has already proved its worth in geophysical and petrophysical analysis and will come to have a long-term impact on many more major areas in geosciences (Fig. 6.1).

Hopefully, ML will provide solutions to some of the fundamental challenges with analyzing geosciences data. One of the future ML solutions will be dealing with multi-scale heterogeneity characterization by integrating data from different sources, resolutions, and formats in one go. The state-of-the-art integration method is an archaic, time-consuming, multi-step process in which debates are still happening on upscaling data and even the definition and procedure of upscaling. Future solutions will be able to generate a tangible product by fusing multi-sensor data at different resolutions, depending on the customer's need for granularity. In a way, ML solutions will become more personalized.

Another area of immediate research need is the automated or semi-automated labeling of data (with interventions) and access to more labeled datasets. Labeling requires domain expertise and time, which impedes the massive deployment of ML in real datasets. More research needs to be conducted on how to generate labels (e.g., facies, fractures, pores, minerals, good/bad data, etc.) using semi-supervision or weak supervision approaches, which can be quality checked later. Most of the recent publications on deep learning deal with classifying features that show clear boundaries on images. These tasks are relatively easy for ML algorithms to recognize. Geologists can also pick these boundaries with more ease before using them as input in supervised classification. However, there are subtle features (e.g., faults with minimal offset, wing cracks, gradational changes in facies, etc.) that may be missed by even an experienced geologist or geophysicist. A clustering algorithm can highlight these subtle features from an entirely data-driven standpoint, enabling us to analyze and label these features properly. Unlike other disciplines, it is somewhat challenging to generate truly benchmark geoscience datasets for broad public use and testing algorithms. There are a limited number of geoscience benchmark datasets available now. New analytics techniques and more collaboration may reduce this problem. In the future, we will see a surge in open and labeled datasets released by the industry and geological surveys. This increase has already started, albeit modestly.

In general, ML has good interpolation capability but not necessarily extrapolation capability. This presents a fundamental barrier to the commercial-scale deployment of ML. In the future, we hope to build better ML models by feeding domain expertise and better features (i.e., physics, chemistry, and geology-based rules) into ML and develop new algorithms and workflows in transfer learning. Physics-informed ML has the potential to simulate the propagation of waves through the subsurface and help us invert recorded waveforms to derive a better picture of the subsurface. We can use these solutions in several sectors, including energy resources, hydrology, storage, construction, and agriculture. Coupled physics and chemistry-informed ML will also help us in predicting fracture geometry, fluid-rock interactions, and reactive transport modeling. Chemistry-informed ML has the potential to predict reaction rates and mineralization, which can be helpful to carbon sequestration and hydrogen storage projects. As these projects, with a focus on reduced carbon emissions, are becoming more common, it is critical to understand the chemical reactions of injected fluids with subsurface rocks and perhaps subsequent mineralization process over time. ML-based solutions can be very helpful to solve such important problems at both laboratory and field scales. Sequence stratigraphy is another specialty waiting for the right ML solution. Feeding specific geology-based rules along with forward stratigraphic models and deep learning models will improve the pace of sequence stratigraphic interpretations and analysis, regardless of the outcrop, seismic, and well log data. With advances in ML, we will also be able to perform better correlations of surface-to-subsurface features, an important research topic in stratigraphy.

Transfer learning has great potential for generating models that can extrapolate, perhaps in baby steps. The current transfer learning models are not helpful for obtaining high-resolution outputs. A combination of physics-based ML with improved transfer learning will be a new direction for ML. Feature engineering will have a role in this research. This will help us explain ML models and understand causalities. Recent analytics tools, such as partial dependence plots, SHAP, and LIME are a modest start to explaining the ML models, but this area needs more research. We can extend the application of physics-informed ML and transfer learning to predict rare events, which are of particular interest in geosciences, for example, earthquakes, volcanic ash beds, bright spots, anomalous pressure and temperature, etc. These examples are all products of fundamental geologic processes such as tectonics, sediment deposition, subsidence, uplift, diagenesis etc. Often, these problems occur in real-time. Because there are a few samples from such events, we need to build benchmark datasets augmented by forward modeling (at least at the regional scale) and envision new techniques that can solve class imbalance problems. For these types of problems, we also need to come up with new metrics for class-specific error analysis. ML can have great future in geosciences if we can solve this problem.

Another area in where ML can make a long-lasting impact is building multi-output classification and regression models. Often, we are interested in deriving multiple rock properties of interest (e.g., poro-perm, fluid saturation, pore pressure, and Poisson's ratio, etc.) simultaneously. We already do this in the simultaneous inversion of seismic data. The same goes for multi-output ML-based classification problems, in which we can classify different facies and faults together. Multi-output dynamic ML models will help us make complex real-time decisions in areas such as time-lapse subsurface monitoring. It is also important because many such properties are conditionally dependent on each other. Generating simultaneous solutions by using stacked ML models can perhaps elucidate their complex relationships and discover new knowledge.

Large companies are already using ML at both research and commercial scales. We also know about intelligent oil fields (aka digital oil fields) in the United States, Kuwait, Saudi Arabia, and other countries. These types of highly instrumented assets and even field laboratories will become more common in the future because these ambitious projects provide integrated solutions to optimizing production and guiding operations safely and efficiently. Companies will have new solutions for automated drilling. Some of these concepts will be impacting the mining and geothermal industries. As many surface mines are exhausted and environmental concerns arise, there will be new research into deep borehole drilling and exploration into mineral resources using ML. Mining engineers have already developed different ML-based tools for mapping 3D orebodies, estimating reserves, designing mines, and simulating mills. Although small companies may not be able to afford all these methods, they can adopt data analytics and ML to solve more mundane problems with feasible goals, including better data digitizing and storage, maintenance, and prediction of missing data that is expensive to acquire in the field.

What lies in the future of ML in the energy industry? Energy is the basis of human civilization. As of 2021, we are undergoing a gradual energy transition. Several companies are attempting to explore new energy resources and adopting new business strategies, including geothermal, solar, wind, rare-earth elements, carbon storage, and hydrogen storage. These new businesses can utilize available data from the existing oilfields (onshore and offshore) to characterize the subsurface—an attractive proposition. This will further accelerate geoscientists' interest in ML and generate resource-specific solutions. ML will have a significant role in integrating energy, economy, and the environment, and geoscientists are poised to be major beneficiaries of that for years to come.

Apart from industry, ML will have a big role in transforming how we teach geosciences to a diverse body of students in academia and how we train new employees. In most of the schools, geoscience is being taught more descriptively and traditionally. As of 2021, it is also true that universities across multiple countries (especially in the western hemisphere) are experiencing a significant downfall in the number of students enrolled in traditional geosciences, which itself is a complex problem and has severe ramifications. Several universities are acting now to encourage students and recruit them to geosciences. We will see more and more quantitative courses being offered, making tomorrow's geoscientists more analytically mature and responsive to society's critical needs. We will see more augmented reality (AR) and virtual reality (VR) used to provide valuable field experiences in remote areas to students who need them. Access to field experience is integral to an effective geoscience education; unfortunately, it is often inaccessible due to logistics, cost, time, and human factors. By adopting new data analytics technologies such as Lidar, drones, and GigaPan, we can provide valuable lessons to students and retain them in geosciences. The same is true is for geophysicists and petrophysicists, who need to see the expressions of different geologic features on seismic and well log data, and other types of remotely sensed data. Instructors will use data analytics to teach affordable online courses, and new software tools will be used in the workplace. Transparent ML solutions may facilitate a more inclusive workplace for the next generation.

One might also wonder that most of the current ML-related research in our field is mere applications, where we borrow the ML algorithms from other fields and directly apply to our problems, but not the other way around. We hardly contribute to other disciplines in ML research. Some of these challenges are due to the nature of our data, including multi-scale geologic heterogeneities, the absence of physics-based rules in most of the available ML models, the lack of labeled data in the public domain, and ever-changing ML algorithms to keep up with. We are also dealing with the problem of converting some of geoscience's fundamental process-based cognitive concepts to mathematical forms (perhaps, we need to coin a new name for this: geo-psychology?).

In this context, it is also important to realize that currently available ML algorithms were invented and implemented to solve non-geosciences problems (e.g., biology, psychology, health sciences, stock exchange, and computer games, etc.). Our datasets are different than other disciplines to an extent. Geologic data is a mixture of multi-scale continuous-to-discrete, regular-to-irregular and image-to-text samples, often filled with missing data (e.g., unconformities and erosional remnants), high-frequency and low-frequency events. In addition, many geologic parameters have a high interdependence or at least conditional dependence, and they are lagged in time or depth. These features preclude a quantitative and comprehensive understanding of the Earth's history and geologic processes and introduce uncertainty to geological interpretations. Future inventions in ML must overcome these fundamental challenges, and perhaps we need to develop new algorithms to suit geosciences data. It may take us some time to develop these fundamental concepts and processes into mathematical forms and models that may transcend across the disciplines and incite new research in the ML community, for example, smart proxy modeling for computational fluid dynamics. In the late 1970's, Peter Vail's work on sequence stratigraphy fundamentally changed the thought-processes in the subsurface geosciences community and the industry. Future research in ML has similar potential.

We should also be careful of what ML can do and what some organizations advertise now. As more and more businesses are picking up ML for the first time without formal training or understanding, there are numerous chances for misadventures, overselling, and an eventual lack of motivation in the long term. It is often our lack of knowledge of a problem, assumptions, principles of ML algorithms, and datasets, not algorithms themselves, which cause ML models to fail. Algorithms are based on solid foundations of mathematics and statistics. At the end of the day, ML is not a snake-oil business. It is a wonderful tool for solving critical and complex problems beyond human capacity. As geoscientists, we need to rise above conventional wisdom and pursue innovative research so we can express fundamental geologic processes in mathematical and statistical forms, based on experiments, observations, and simulations, and assist in making data-driven decisions to make us more successful. ML could be an enabler of this grand endeavor!

Printed in the United States
by Baker & Taylor Publisher Services